Networks, Alliances and Partnerships in the Innovation Process

Economics of Science, Technology and Innovation
VOLUME 28

The titles published in this series are listed at the end of this volume.
Economics of Science, Technology and Innovation

Networks, Alliances and Partnerships in the Innovation Process

Edited by

JOHN DE LA MOTHE
PRIME, University of Ottawa

and

ALBERT N. LINK
Department of Economics, University of North Carolina at Greensboro

KLUWER ACADEMIC PUBLISHERS
Boston / Dordrecht / London

Distributors for North, Central and South America:
Kluwer Academic Publishers
101 Philip Drive
Assinippi Park
Norwell, Massachusetts 02061 USA
Telephone (781) 871-6600
Fax (781) 681-9045
E-Mail: kluwer@wkap.com

Distributors for all other countries:
Kluwer Academic Publishers Group
Post Office Box 322
3300 AH Dordrecht, THE NETHERLANDS
Telephone 31 786 576 000
Fax 31 786 576 474
E-Mail: services@wkap.nl

Electronic Services < http://www.wkap.nl>

Library of Congress Cataloging-in-Publication Data

A C.I.P. Catalogue record for this book is available
from the Library of Congress.

NETWORKS, ALLIANCES AND PARTNERSHIPS IN
THE INNOVATION PROCESS

John de la Mothe and Albert N. Link (eds.)

ISBN: 1-4020-7172-8

Printed on acid-free paper.

Printed in the United States of America.

CONTENTS

PART III: PRACTICE

PART IV: IMPACTS

PART IV: CONCLUSION

CONTRIBUTORS

Frances Anderson, SIEID, Statistics Canada.

Anthony Arundel, MERIT, University of Maastricht.

Robin N. Auger, Centre for Science and Technology Policy, School of Public Policy, George Mason University.

Catalina Bordoy, MERIT, University of Maastricht.

John de la Mothe, PRIME, School of Management, University of Ottawa.

Dominique Foray, CERI, OECD, Paris.

John Jankowski, Research & Development Statistics Program, The National Science Foundation.

Don E. Kash, Centre for Science and Technology Policy, School of Public Policy, George Mason University.

Ning Li, Centre for Science and Technology Policy, School of Public Policy, George Mason University.

Albert N. Link, Department of Economics, University of North Carolina at Greensboro.

Jacques Lyrette, Technology and Industry Support, National Research Council of Canada.

Vincent Mangematin, INRA, Université Pierre Mendès-France.

Susan McDaniel, Department of Sociology, University of Alberta.

Chuck McNiven, SIEID, Statistics Canada.

Francisco Moris, Research & Development Statistics Program, The National Science Foundation.

Lionel Nesta, INRA, Université Pierre Mendès-France.

Paul Quintas, The Business School, The Open University.

Antoine Rose, SIEID, Statistics Canada.

W. Edward Steinmueller, SPRU, University of Sussex.

Graham Vickery, Information, Computer and Communications Policy Division, Directorate for Science, Technology and Industry (DSTI), OECD.

ACKNOWLEDGEMENTS

This volume is one outcome of a fifth Statistics Canada / PRIME joint research workshop that was held in Ottawa Canada during March 2002 on the general theme of Networks, Alliances and Partnerships in the Innovation Process. This five year project began in March 1997 and resulted in the 1998 book, also published by Kluwer in this series, that we co-edited on Local and Regional Systems of Innovation. Subsequent volumes have focused on Information, Innovation and Impacts (1999), The Economic and Social Dynamics of Biotechnology (2000), and Knowledge Management in the Innovation Process (2001).

In bringing this final volume in this phase of the Innovation Dynamics Project together, from the idea and the workshop to this book, we would very much like to thank:

- Dr. Fred Gault of Statistics Canada for supporting the project and for collaborating with us in its development and delivery;
- the National Science Foundation – particularly Lynda Carlson – for its sponsorship;
- several students from PRIME for their help in making the meeting a reality, particularly Tyler Chamberlin and Jennifer Garrett;
- Tom Randall of Kluwer for once again guiding and encouraging this project so kindly;
- Louise Earl of Statistics Canada for significant logistical and editorial support;
- Bo Carlson and Christiano Antonelli for again inviting this book into their fine series; and
- to each of the contributors – both those who were at the workshop and those who are represented in this volume – for their very stimulating contributions.

JdlM/AL

PREFACE

This is the fifth book from a project that examines technological change from economic, social and statistical perspectives. The previous four books dealt with innovation, the first with its regional aspects, the second and third with the impacts of two transforming technologies (the information and communication technologies (ICTs) and biotechnology), and the fourth with the practices of knowledge management. The present volume continues this project with a focus on alliances, networks and partnerships in the innovation process.

This book, and the related workshop that gave rise to it, were challenging undertakings because the subject was not a set of technologies, or practices, and it was not tied together by geography as were papers in the first book. Alliances, networks and partnerships transcend geography and subsume technologies and practices. Understanding their role in the process of innovation is complex and it is complex because the object of study is the linkage between actors in a system and the linkages, which are the structure of the system, change over time and in response to economic and social signals. Any attempt to understand the role of linkages in the innovation process cannot happen without reference to the actors and their activities, as well as their linkages.

The previous books were also clearly linked to policy initiatives, actual or potential, such as the promotion of regional innovation, which these days would be cast in terms of incubating technology clusters, or the promotion of the use or planned use of technologies and practices. With alliances, networks and partnerships there is no single committee of the Organization for Economic Co-operation and Development (OECD) preoccupied with their understanding or promotion. It is the nature of linkages to cross boundaries and the contributions from the OECD to the workshop came from quite different parts of the organization. While there is an extensive literature on research alliances, there is less available on linkages and the innovation process, which is where this volume is situated. Of course, the linkages considered here are not just those between firms, but also with governments and with higher education.

Linkages change with time and with technology. In May 1990 in Montreal there was conference, organized by Christian DeBresson, on Networks of Innovators and Chris Freeman, in his summary of that conference (Research Policy, Vol. 20, no. 5, (1991) 499-514) made the point that many of the characteristics of innovative firms had been known for some decades. He provided a list from a paper published in 1972 which remains applicable thirty years later. However, he did note that networking in the 90s would be different and he went on to observe that:

> IT (information technology) not only greatly facilitates various forms of networking, but has inherent characteristics, such as rapid change in design, customisation, flexibility and so forth, which, together with its systemic nature and the variety and complexity of applications, will lead to a permanent shift of industrial structure and behavior. This will assign to networking a greatly enhanced role in the future.

During the 90s, technology did change. Knowledge workers all gained a PC and learned to use the Internet and then the web. More knowledge was codified and made available within the organization and on the web and the tacit knowledge moved higher up the value chain. The technology helped to flatten organizational hierarchies and it made it easier for firms to align themselves, to use networks, and to enter into partnerships in order to develop new or improved processes or to introduce new or radically improved products to the market.

The changes in the 90s and the current interest in the subject of how firms engage in alliances, networks and partnerships, across the world, along the supply chain, and in local clusters, provided the motivation for the 2002 research workshop. The 2002 workshop also built on a US National Science Foundation (NSF) workshop on research partnerships held in Washington in October 2000, and reported on in Strategic Partnerships: Proceedings from an NSF Workshop (NSF, 2001). The interest of the US NSF in going beyond research partnerships to look at the influence of linkages on innovation in the firm gave rise to a productive collaboration as well as the material support which made the workshop possible.

As with previous workshops, the day began with two presentations designed to frame the discussion which followed during the day, and the resulting papers by Al Link and Ed Steinmueller follow the introductory chapter in this book. In keeping with the subject, both provided links to other activities, Al Link to the related NSF workshop, and Ed Steinmueller to the Science and Technology Policy Research Unit (SPRU) at the University of Sussex which is home to Chris Freeman. The section on frameworks is followed by one on statistical measurement which is designed to show what aspects of the subject are amenable, now, to statistical measurement and analysis. This and the following section on the practice of building linkages provide a reality check both for the framework papers and for the papers on economic and social impact which follow the section on practice.

All of the authors in this book exemplify alliances, networks and partnerships and none more than John de la Mothe, who has been the common element in all five volumes and in all of the Statistics Canada / PRIME research workshops.

Fred Gault
Director, Science, Innovation and Electronic Information Division
Statistics Canada

PART I

INTRODUCTION AND FRAMEWORKS

Chapter One
INTRODUCTION

John de la Mothe
Canada Research Chair in Innovation Strategy,
PRIME, University of Ottawa, &
Visiting Professor of Science and Technology Studies,
University College London

and

Albert N. Link
Department of Economics
University of North Carolina at Greensboro

Why should we be concerned about networks, alliances, and partnerships in the innovation process, now, in the 21st century? After all, there have always been such linkages in the history of human social and economic activity and they have always played a role in bringing new products to market, in creating new processes or in developing new services. So what makes this an important topic for current research?

Three reasons drive this investigation, each of which has significant import for economic and social analysis, for policy making, and for statistical measurement. These reasons are the confluence of globalization, the growing importance of knowledge flows, and the changing management of firms. Each of these has shifted over the past decade and has changed the ways in which networks, alliances, and partnerships contribute to the innovation process. Underpinning these activities is the information and communication technology (ICT) infrastructure.

A decade ago we did not have the world-wide web or the ubiquity of the Internet. Now, thanks to changes in regulation, technology and software applications, we have falling prices for global communications, access to data, information and codified knowledge, and the opportunity to trade electronically in goods and services. This connects economic agents globally, and locally, and it connects economies at quite different stages of economic development.

Globalization puts downward pressure on commodity prices and on labor costs as traders can find their goods in the best markets. Software designers in Europe or North America, for example, can contract for design services with firms in Bangalore. ICTs, and the applications that run on the ICT infrastructure, help to make this happen and they also facilitate knowledge flows - knowledge of markets, of clients, of suppliers, of production processes, and of management standards and practices. The ability to share knowledge within an organization, public or private, has led to a flattening of the management hierarchy, which has implications for the way workers function and firms compete.

Workers in firms with flat management structures have to be more autonomous to function and they have to be open to networking, to striking alliances and engaging in partnerships in order to achieve their objectives. The same is true of the orgnaization itself and the flexibility of communication allows organizations to align themselves, to form geographically anchored clusters, or to build partnerships with other organizations anywhere on the planet to support the value chain of which they are a part.

The interactions amongst people build social capital, based on trust, which is a key factor in 21st century production and so is the 'network capital' which firms build in their linkages with other institutions. For people to be more autonomous, they have to invest over their lifetime in building and maintaining their human capital. The extension of the production and management processes over space and time change the approach of economic actors to the value chain and to the product life cycle. All of these have impact on how business is done and products are traded .

The present state of the private sector, in industrialized countries, can therefore be characterized as 'alliance capitalism', the successor of 'hierarchical capitalism' which dominated the period from about 1875 to 1980, and 'entrepreneurial capitalism' which reigned through the first industrial revolution and the latter half of the scientific revolution. This evolution has paralleled accelerating economic, social and technological changes which have brought challenges to the sovereignty of nations through alliances of multinational firms, currency traders, and the electronic delivery of cultural products. Of course, all of this is done to bring to market new products and to introduce new processes for their creation and for their delivery. All of this involves networks, alliances, and partnerships in the innovation process. Their study is thus of immediate relevance.

To advance our understanding and thinking concerning these issues, this book brings together reflections on conceptual frameworks, statistical measurement, practices and impacts of networks, alliances, and partnerships in the innovative process and it addresses questions that are arising in the public and private sector about how these linkages function in a complex and dynamic economic and social environment, as well has the impact of these linkages.

In the two chapters offering insight into frameworks, Albert Link and Edward Steinmueller explore the differing rationale that motivate private and pulic sector strategies. Link first discusses the entrepreneurial process in the context of growth economics with reference to infrastructure technology, research partnerships and the science base. His explanation of a variety of small business initiatives and research consortia amplifies his core points, including dual-use programs Edward Steinmueller extends the conversation by focusing in detail upon rationale, indicators and significance of collaborative innovation. To do so he studies the context of research, re-examining the theoretical foundations of research networks including divisions of labor and coordination functions.

In the two chapters on measurement, the authors (Anderson, McNiven and Rose; and Arundel and Bordoy) discuss differences in manufacturing and biotechnology by using the Innovation Survey (1999) of Statistics Canada. The first authors offer results from this survey and look at firm size, internal R&D, stage of development, type of partners, purpose of collaboration by size and by stage of development. Anthony Arundel and Catalina Bordoy continue the measurement assessment by looking at in-house and ex-house differences as sources for innovative knowledge. They use the KNOW survey that, in 2000, looked at seven EU countries.

The 'practice' section of this volume highlights the experience of Jacques Lyrette - from the National Research Council of Canada and the other - by Francisco Moris and John Jankowski - from the U.S. National Science Foundation. This is followed by a detailed set of questionings by Paul Quintas concerning the division of knowledge. This examination flows from a series of statements: examples of which are that knowledge cannot be regarded as a commodity that is easily traded or shared; that knowledge is created through a series of dynamic processes in specific contexts; and, that knowledge has important social dimensions. In his discussion, Quintas looks at extra-organizational knowledge as a prelude to inspecting transient and persistent networks, and their linkages to innovative behavior. This section closes with some evidence from case studies.

In the final substantive section of the book, six authors offer varying perspectives on the internal dynamics and importance of networks, alliances and partnerships. Kash, Auger and Li begin by discussing the organizational requirements for innovation in complex technologies. This approach in itself is portentous given that many analyses hitherto prefer to simplify frameworks through the initial statement *ceteris peribus* – 'all things being equal' (which they never are) - and then to, for example, count the number of firms, the number of patents, the number of alliances, and then to draw undifferentiated conclusions. The Kash, Auger and Li analysis is much richer and truer to real organizations, offering real avenues for researchers, statisticians and policy makers. Dominique Foray continues this line of assessment by looking at the role of networks and partnerships as an endogenous mechanism to reduce dynamic transaction costs. Lionel Nesta and Vincent Mangematin excavate the literature on industry life cycles in order to plumb its utility or applicability to knowledge generation in technology-intensive networks. What they suggest (and this links to and amplifies the offerings of, for example, Kash *et al.* and Quintas, in this volume) is indeed interesting. While technologies and networks throw up difficult issues of complexity, there may be underlying and more continuous cycles of behaviour in business, investment and economic activity. Graham Vickery extends and shifts the impact section of the book to examine the restructuring of value chains by reviewing evidence on the impacts of the Internet. This chapter draws in part on the OECD E-Commerce Business Impacts Project. One of the several advances noted in this chapter is to go beyond the traditional product innovation and process innovation and to develop the notion of relational and organizational innovation. Susan McDaniel again shifts the framework of the volume by moving well beyond micro-social contexts that are popular in sociological analysis in order to look at the innovation potentiality of social environments which she describes, using evolutionary imagery, as 'a waltz'. The final chapter in this section is by John de la Mothe who looks at networks in policy making and their changing roles in innovation systems. After critiquing several streams of network literature, from the Schumpeterian and Kondratiev innovation literatures to the less structured and less evocative public management literatures, he then reviews some examples of policy-initiated research networks between the public and private sectors in what he describes as 'adaptive' public policies.

Taken together we believe that policy makers, scholars and statisticians can learn much from this volume about the growing importance of networks, alliances and partnerships in the innovation process. In so doing we attempt to deepen our appreciation for the complex relationships that animate knowledge, learning and information in the new economy. In so doing we suggest paths for analysis, measurement and better decision making.

Chapter Two
PRIVATE-SECTOR AND PUBLIC-SECTOR STRATEGIES TO ENCOURAGE TECHNOLOGICAL ALLIANCES

Albert N. Link
Department of Economics
University of North Carolina at Greensboro

1. INTRODUCTION

'Technological alliances,' 'research joint ventures,' 'strategic networks' are all terms that have become commonplace, as well as synonymous, within the so-called science and technology policy literature; however, few have attempted to distinguish among the meaning of these terms, much less to systematically conceptualize precisely to what they refer. For nearly two decades, innovative agents, firms specifically, have formally gone outside of their organizational boundaries to acquire technological knowledge from others in an effort to reduce innovation costs, to decrease time to market, and/or to leap-frog existing technologies. Whatever the strategic motive, and innovation is indeed a purposeful strategic action, these alliances have emphasized through example the importance of knowledge flows to on-going research agendas.[1] Thus, one might think of 'technological alliances' as an arrangement—formal or informal—that comes about—intentionally, culturally, or by accident—for the purpose of facilitating knowledge flows.

In this chapter I offer one possible framework for conceptualizing technological alliances, both as a strategic part of a firm's innovative behavior and as a policy instrument. This framework illustrates that there are both public- and private-sector strategies at work to facilitate the formation of such alliances. Public-sector strategies are often manifested in the form of legislative initiatives, while private-sector strategies are either a response to these initiatives or captive strategic efforts. In either case, the framework distinguishes among them while focusing their collective intent to enhance firm innovation.

After presenting the framework, evidence about alliances in the United States is offered. This national focus should not be interpreted as delimiting the usefulness of the framework for more general policy use; rather, U.S. illustrations are offered because of greater expertise in that domain. The chapter concludes with a brief discussion of the accountability responsibilities that go hand-in-hand with these technology-based initiatives.

2. THE ENTREPRENEURIAL PROCESS

From an economic perspective, the concept of entrepreneurial innovation can be traced back to the Physiocrats in France in the mid-1700s. Baudeau (1910, p. 46), for example, referred to a process guided by an active agent, which he called an entrepreneur, within a capitalistic system:[2]

> Such is the goal of the grand productive enterprises: first to increase the harvest by two, three, four, ten times if possible; secondly to reduce the amount of labor employed and so reduce costs by a half, a third, a fourth, or a tenth, whatever possible.

Embedded in this conceptualization of entrepreneurship is the notion of an innovative process, one perhaps as simple as the perception of new technology to be adopted from others so as to increase agricultural yield, or one as refined as the actual development of a new technology to do the same. When the process is completed, and when the innovation is put into use, there will be productivity gains.

Entrepreneurship is thus a process, with its output being the promotion of one's own innovation or the adoption of another's innovation. The term 'entrepreneurship' is commonly used to refer to a businessman or even to a risk taker. I am using the term in a much broader sense; an entrepreneur is one who perceives an opportunity and has the ability to act upon it. Hence, entrepreneurship is a process that involves both *perception* and *action*. The perception of the opportunity may be influenced by changes in strategic directions or competitive markets, but perception of the opportunity is the fundamental first step.[3] The consequent step is the ability to act on that perception. What defines the entrepreneur is the ability to move technology forward into innovation. The technology may be discovered or developed by others. The entrepreneur is able to recognize the commercial potential of the invention and organize the capital, talent, and other resources that turn an invention into a commercially viable innovation.

What are the requisite resources needed for action, which takes the perception of an opportunity forward to result in an innovation? One obvious answer is *research and development (R&D)*, that is, the commitment of resources to invention and innovation. R&D not only provides a stock of knowledge to encourage perception but also the ability for the firm to foster action. However, firms that do not conduct R&D can still be entrepreneurial, as discussed above. In such firms, innovations are likely to be introduced rather than produced. Such firms act in an entrepreneurial manner by hiring creative individuals and providing them with an environment conducive for the blossoming of their talents.

Consider R&D-active firms. The R&D they conduct serves two general purposes. First, it provides the resource base from which the firm can respond to a potential opportunity with perceived strategic merit or to a known technical opportunity that allows the firm to develop a commercial market. Second, those scientists involved in R&D are the internal resource that facilitates the firm's being able to make decisions regarding the technical merits of others' innovations and how effectively those innovations will interface with its existing technological environment. The firm may choose to purchase or license technology or undertake a new R&D endeavor. In this latter sense, one important role of internal or in-house R&D is to enhance the absorptive capacity of the firm.

Thus, the role of R&D in enhancing the absorptive capacity of the firm goes beyond simply assessing the technical merits of potentially purchasable technology. It allows the firm to interpret the extant technical literature, to interface when necessary with the research laboratories of others, in a research partnership relationship; or to acquire technical explanations from, say, a federal laboratory or university laboratory; or simply to solve internal technical problems.

Figure 1 provides an initial analysis of what I term the entrepreneurial process. This initial look will be expanded upon below, but here it introduces the following three concepts:

1) The organization, typically a private-sector firm, has a focus that results from it's agreed upon strategic direction. This strategic direction, coupled with competitive market conditions, generates an entrepreneurial response.
2) The purposive activity associated with the entrepreneurial response leads to an innovation.
3) There are market forces at work that are, in part, beyond the influence of the firm and these forces determine the economic value of the innovation and hence the value added to the project as well as to the user of the innovation.

Figure 1. The Entrepreneurial Process: An Initial Look

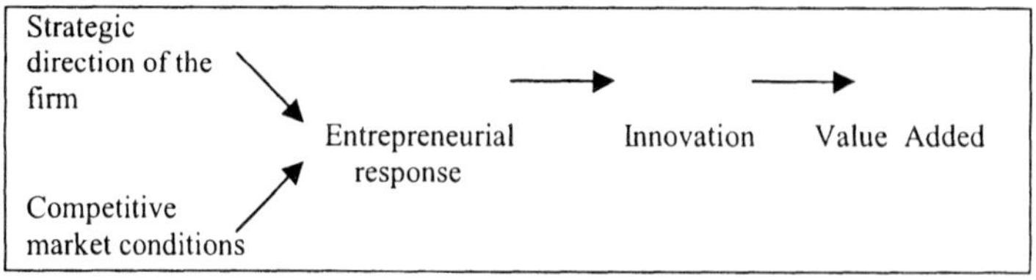

Figure 1 can be expanded on the basis of what is frequently called 'old growth theory' and 'new growth theory.' The early literature on economic growth is formulated in terms of a production function. If output, Q, can be defined in terms of the two most basic factors of production, the stock of capital, K, and the stock of labor, L, then a basic production function can be written as:

(1) $Q = f(K, L)$.

Equation (1) denotes that the firm's, industry's, or economy's output will change in response to changes in either the quantity or quality of capital or the quantity or quality of labor. However, to account for other influences on output such as new technology, the production relationship in equation (1) can be modified most simply to include a catch-all variable, A, as:

(2) $Q = A\,F(K, L)$

where, to be more specific, A is a shift factor to account for exogenous technological factors (as opposed to conventional factor inputs such as K and L) that affect production, and where, because of the inclusion of A in equation (2), $F(K,L)$ is distinct from $f(K,L)$ in equation (1).

By dividing both sides of equation (2) by the combination of K and L inputs (i.e., by total factors) denoted by $F(K,L)$, variable A can be interpreted as an index of output per unit of all inputs or of total factor productivity:

(3) $A = Q / F(K, L)$

Changes in A from equation (3) measure what is called total factor productivity growth, or technological advancement.

The academic literature is replete with theories to explain growth over time. The so-called 'old growth theory' literature is based on more sophisticated versions of equation (1). That is, this literature emphasizes additional inputs aside from K and L such as investments in R&D and education.[4] As well, it emphasizes the greater specificity by which inputs are measured including consideration for the heterogeneity of K and L (e.g., new vintages of K that embody others' technological investments).

The so-called 'new growth theory' emphasizes the influence of other factors on growth that are not directly specified in an expanded version of equation (1). These factors include, for example, technologies or efficiencies that spillover into a firm's production function either from other firms or from general advances in the economy (such as information technology) or that spillover into a nation's production function from, say, trade policies. New growth theory is also based on careful, explicit analytical modeling of the incentives of agents to invest in new technology.

Figure 2 expands Figure 1 to incorporate these ideas. First, in-house or private investment in R&D is shown to have an influence on how the firm responds to the interaction of its strategic direction and its competitive environment. Second, I have capsulated the essence of new growth theory by simply acknowledging that external influences affect firm performance directly as well as indirectly through innovation.

Regardless of whether one adheres to the more narrow old theories or the broader new theories, the evidence is overwhelming that technology drives economic growth. There was renewed interest in promoting economic growth in the postwar aftermath of the destruction of the industrial base of many nations. Thus, it is not surprising that greater attention was devoted to the analysis of R&D, and it is likewise not surprising that policy initiatives have focused on R&D.

Figure 2. The Entrepreneurial Process: A Second Look

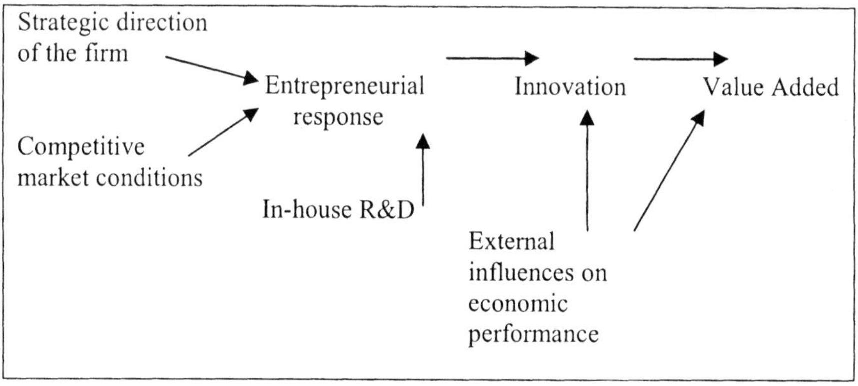

Figure 2 can be expanded upon yet again in Figure 3 to incorporate additional technology elements. The inclusion of these elements is critical to understanding technology alliances, and to an understanding of how they interact with in-house R&D and thus ultimately with innovation.

The first technology element is what is called *infrastructure technology*. It originates in federal laboratories, such as the National Institute of Standards and Technology (NIST) in the United States, or from the environment created for firms from being located in, say, a science park. The second technology element comes from involvement in *research partnerships*, with other firms or perhaps with either a university or a federal laboratory. Finally, the *science base* also influences the level of in-house R&D activity. The science base conceptualizes the stock of knowledge generated from basic research, thus residing in the public domain—and the public domain is international in scope—generally in the form of scientific journals but also it is in part embodied in university scientists.

Two internal feedback mechanisms, depicted by dashed lines, are also added to Figure 3. The first feedback flows from innovation to the science base. Once an innovation exists, knowledge has been created and it too will reside in the public domain. The second feedback extends from innovation to competitive market conditions. It reflects the extent to which innovation can alter the competitive landscape. This can be seen, for example, in the evolution of industries, where new technologies eclipse old. Although profit opportunities create an incentive for innovation, innovation can subsequently alter the structure of a market and in so doing, the profit opportunities in that marketplace. Generally speaking, market structure and market behavior are jointly endogenous. More recent literature on industry structure, applied to issues such as innovation, often uses a game-theory approach to separate out endogenous effects from true exogenous conditions.

A forward-looking entrepreneurial strategy takes seriously the feedback from innovation to market conditions. Such strategy shapes the nature of rivalry in the innovation process. For example, in one scenario, a significant innovation can create a monopoly for the successful entrepreneur, either due to a patent award or from the failure of rivals to quickly imitate. In this case, absent other market failures such as spillovers, the profit incentive likely leads to intense competition between competing entrepreneurs in their race to innovate. It follows that the initial competitive landscape ultimately changes to a monopoly. Interestingly, game theoretical models shows that excessive duplicative R&D may occur in this scenario. However, as discussed in previous sections, and confirmed by game theoretical models, an opposite result can occur if significant spillovers of knowledge flow from one firm to another. That is, strategic R&D competition may yield too little innovative activity because the flow of knowledge to competitors has a negative effect on the entrepreneur's position.

Figure 3 is not only a useful summary device for denoting sources—technology alliances—through which private-sector firms acquire technical knowledge, but it is also a useful device for summarizing any industrialized nation's public-sector policies toward innovation, acting through in-house R&D. The patent system and tax incentives provide direct incentives to the firm to increase its level of R&D. Initiatives to encourage collaborative (firm with firm) research effect the efficiency of in-house R&D by reducing duplicative research costs and shortening the fundamental research stage. Government-provided infrastructure technology, through standards or test methods, for example, reduces transaction costs in the market, thus lowering the marginal cost of R&D. Finally, direct federal support of university research continually enriches the science base thus facilitating the in-house R&D process.

Figure 3. The Entrepreneurial Process: An Integrated Look

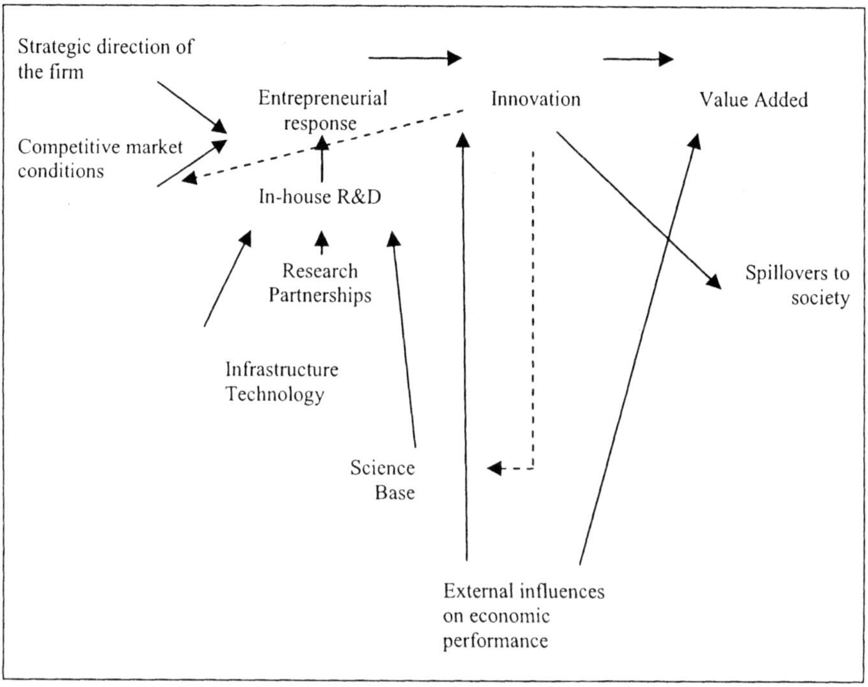

3. PUBLIC/PRIVATE PARTNERSHIPS

In a general sense, all three of the technology elements mentioned above and noted in Figure 3 are what may be called public/private partnerships, where the term 'partnership' is used in this chapter to refer to any innovation-based relationship that involves 'public' and 'private' resources, and where 'resources' refer to financial resources, research resources, and infrastructure resources.[5]

There are a number of useful ways to categorize public/private partnerships. For example, cooperative technology programs or partnerships can be classified in terms of the benefits and services they offer to industry. Toward that end, five so-called functional categories have been posited:[6]

1) Technology Development: research and applications for new or enhanced industrial products and processes
2) Industrial Problem Solving: identifying and resolving company-level industrial needs through technology and best-practice applications
3) Technology Financing: public capital or help in gaining access to private capital
4) Start-up Assistance: aid to new small technology-based businesses
5) Teaming: help in forming strategic partnerships and alliances.

In this chapter, public/private partnerships are alternatively and uniquely classified in terms of the economic objective of the partnership. Public/private partnerships can be formed to accomplish at least three broad, although not necessarily independent, economic objectives:

1) Public/private partnerships can exist to leverage the social benefits associated with federal R&D activity.
2) Public/private partnerships can exit to enhance the competitive position of industry in the global marketplace.
3) Public private partnerships can exist to leverage industrial R&D to meet military or defense needs.

Table 1 posits a taxonomy of public/private partnerships in terms of the economic objective for creating the partnership as well as the public resources marshaled to facilitate the objectives of the partnership. Not surprising, because public/private partnerships are a direct funding policy mechanism, the public sector brings financial resources to the relationship. But as well, the history of partnerships in the United States reveals that infrastructural resources—changes in the research climate or environment (e.g., a change in the antitrust environment which thereby encourages the formation of research joint ventures as an infrastructural resource) in which the partnership exists—are also provided by the government. It should be emphasized, as also noted in Table 1, that public research resources are only an integral part of the public resource mix brought to a public/private partnership that leverage federal R&D. The absence of this characteristic in the other two partnership areas should not be interpreted to suggest that the research undertaken in federal laboratories, including federally-sponsored university laboratories for example, does not have an important economic objective and does not spillover to enhance industry. Rather, it is extremely important research, and it enhances the efficiency with which the private sector conducts its applied research and development. Such public research is simply not the most fundamental aspect of public resources that are brought into the types of public/private partnerships.

Table 1. Taxonomy of Public/Private Partnerships

	Economic Objective		
Aspect of Direct Funding by the Public Sector	Leverage Federal R&D	Enhance Industrial Competitiveness	Leverage Industrial R&D
Financial Resources	yes	yes	yes
Research Resources	yes	no	no
Infrastructural Resources	yes	yes	yes

4. INSTITUTIONAL HISTORY OF PUBLIC/PRIVATE PARTNERSHIPS IN THE UNITED STATES

One can trace the origins of a broadly-defined cooperative technology development policy, or of technology partnerships, in the United States at least to the Lincoln Administration. In 1862, the Morrill Act established what was known as the land-grant college system. The Act created a partnership between the federal and state governments to cooperate with the private sector in technology development. That Act charged states to develop colleges to offer curricula in agricultural and mechanical arts. Then, in 1887, the Hatch Act provided resources for a system of state agricultural experiment stations that would be under the auspices of land-grant colleges and universities. A partnership among the various levels of government was established by the Smith-Lever Act of 1914. The Cooperative Agricultural Extension Service was charged to deliver the practical benefits of research to citizens though an extension service.

According to Carr (1995, p.11):

> Until the end of the 1970s, the philosophy behind the dissemination of federally-funded research was that if the public paid for the research, the resulting intellectual property should be made equally available to all interested parties.

The 1970s and 1980s witnessed many foreign competitors begin to successfully challenge the long-standing dominance of the United States not only in world markets but also in the domestic market. It soon became clear to public-sector and private-sector policy leaders that a change in its philosophy of federal R&D support was needed; traditional processes of public/private technology development needed to be re-examined. The Office of Technology Policy (1996) puts this change in public policy mind set into perspective, and it advanced three important reasons for such a re-examination of public policy:

1) Global competitors were better able to appropriate the output of U.S. basic and mission-oriented research as their technical sophistication grew.
2) Traditional public sector mechanisms of technology development, and transfer, took too long in an era of accelerating private sector development.
3) U.S. government R&D represented a declining share of world R&D as globally competing nations increased their public funding; hence, the marginal benefits to industry for additional public moneys (allocated in the same manner as historically) declined.

Stated alternatively, but maintaining the same general theme, Link and Tassey (1987, p. 4, 131) reflect on what were then (early 1980s) recent changes in the competitive environment of U.S. industry:

> Today, there is a new order of competition in the world. An inescapable element of the competition is technology ... With the advent of technology-based economies [throughout the world,] the increase in the number of world competitors has been greater than the increase in the size of the world market. What has resulted from this is a significant shortening of technology life cycles ... As such, effective long-run competitive strategies will have to deal explicitly with technology ... [C]ompetitive survival will depend on technology-based strategies. These strategies will have to evolve from new philosophies about interdependence ... The importance of interdependence arises from the need of a domestic industry to rapidly and efficiently develop complex technological elements from which specific applications (innovations) are drawn for competitive activity ... [Accordingly], government must expand and adapt its role ... with industry for more effective joint planning in research.

Beginning with legislation in 1980, as summarized only briefly in this chapter, a new era in U.S. technology policy began. This new era was based on the belief that the global competitiveness of U.S. firms can be enhanced through legislation to bolster the commercial impact of federal R&D investments. As such, using the terminology of the Office of Technology Policy (1996, p. 26), "a new paradigm for public/private technology partnerships emerged." Prior to the 1980s, the federal government viewed itself, as did industry, as a customer for industry's technology development programs. Beginning in the early 1980s that view changed, and the federal government began to view itself as a partner with industry in joint technology development programs, although its collaboration was primarily in terms of financial rather than research resource support.

The first major piece of federal legislation that was intended to leverage public/private technology development within this new paradigm was the Stevenson-Wydler Technology Innovation Act of 1980 (P.L. 96-480). This Act was predicated on the premise that federal laboratories embody important and industrially-useful technology. Accordingly, each federal laboratory was mandated to establish an Office of Research and Technology Application (OTRA) to facilitate the transfer public technology to the private sector. Such facilitation was part of the authority given to the Department of Commerce under the Act, namely:

> ...to enhance technological innovation for commercial and public purposes ... including a strong national policy supporting domestic technology transfer and utilization of science and technology resources of the federal government.

The University and Small Business Patent Procedure Act of 1980 (P.L. 96-517), also known as the Bayh-Dole Act, reformed federal patent policy thereby providing increased incentives for the diffusion of federally-funded innovation results. More specifically, universities, non-profit organizations, and small businesses were permitted to obtain title to innovations they developed with the use of governmental financial support, and the Act allowed federal agencies to grant exclusive licenses to their technology to industry.

In a broad sense, Stevenson-Wydler and Bayh-Dole institutionalized public/private partnerships through the more efficient transfer of federally-developed technology into the private sector. This represents a form of public infrastructural support to industry. The Small Business Innovation Development Act of 1982 (P.L. 97-219) required that federal agencies provide special funds to support small business R&D that complemented the funding agency's mission. This was called the Small Business Innovation Research (SBIR) program. Based on the premise, as stated in the Act, that "small business is the principle source of significant innovation in the Nation," and small businesses are "among the most cost-effective performers of research and development and are particularly capable of developing research and development results into new products," the Act lists its purposes for, among other things, establishing the SBIR program:

1) To stimulate technological innovation
2) To use small businesses to meet federal research and development needs
3) To foster and encourage participation by minority and disadvantaged persons in technological innovation; and

4) To increase private sector commercialized innovations derived from federal research and development.

The Small Business Innovation Research Program Reauthorization Act of 1992 gave reauthorization to the SBIR program because the program has "effectively stimulated the commercialization of technology development through federal research and development, benefiting both the public and private sectors of the Nation."

The National Cooperative Research Act (NCRA) of 1984 (P.L. 98-462) has two main objectives. The first objective was to establish a rule of reason for evaluating the antitrust implications of research joint ventures (RJVs) on an individual case basis and to limit potential liability to actual rather than treble damages as is more common under antitrust law. The NCRA was amended by the National Cooperative Research and Production Act (NCRPA) of 1993 (P.L. 103-42). The purpose of this amendment was to broaden the scope of the original NCRA to include production activities as well as research in general.

The Trademark Clarification Act of 1984 (P.L. 98-620) set forth new licensing and royalty regulations thereby taking technology from federally-funded facilities into the private sector. Specifically, the Act permitted government-owned, contractor-operated (GOCO) laboratories to make decisions regarding which patents to license to the private sector, and contractors could receive royalties on such patents.

The Federal Technology Transfer Act of 1986 (P.L. 99-502), the last major piece of federal legislation designed to leverage the economic impact of federal investments in R&D, amended the Stevenson-Wydler Act (Office of Technology Policy, 1996). Among other things, the Act made explicit that technology transfer was the responsibility of all federal laboratory scientists and engineers. Specifically, it authorized cooperative research and development agreements (CRADAs) to be established between public/private research organizations.

This Act was later amended by the National Competitiveness Technology Transfer Act of 1989. The 1989 Act expanded the definition of federal laboratory from one that is government-owned and government-operated (GOGO) to one that is a GOCO. The implication of this Act is to allow GOCOs to enter into CRADAs.

The Omnibus Trade and Competitiveness Act of 1988 (P.L. 100-418) not only changed the name of the National Bureau of Standards to the National Institute of Standards and Technology, but also facilitated the ability of Congress to enact two so-called direct competitiveness programs—the Advanced Technology Program (ATP) and the Manufacturing Extension Partnership (MEP). The MEP is a partnership in name more than in research mission. MEPs assist small-and medium-sized manufacturers assess their technology needs and then facilitate its purchase and implementation. The

American Technology Preeminence Act of 1991 (P.L. 102-245) later clarified the mission of the ATP.

The goals of the ATP, as stated in its enabling legislation, the Omnibus Trade and Competitiveness Act of 1988, and modified by the American Technology Preeminence Act of 1991, are to assist U.S. business in creating and applying the generic technology and research results necessary to:

1) Commercialize significant new scientific discoveries and technologies rapidly; and
2) Refine manufacturing technologies.

As restated in the *Federal Register* on July 24, 1990:

> ...the ATP ... will assist U.S. businesses to improve their competitive position and promote U.S. economic growth by accelerating the development of a variety of pre-competitive generic technologies by means of grants and cooperative agreements.

Finally, the Defense Conversion, Reinvestment, and Transition Assistance Act of 1992 (P.L. 102-484) created an infrastructure for dual-use partnerships. Through Technology Reinvestment Project (TRP) partnerships, the Department of Defense was given the ability to leverage the potential advantages of advanced commercial technologies to meet the Department's needs. TRP partnerships speed the development of a commercial technology so that a domestic market that is self sustaining, will develop itself sooner than would otherwise be expected; TRP partnerships then focus the identified commercial technology development in a way that simultaneously (e.g., dual use) meets the needs of industry and military. TRP partnerships benefit the defense sector first and foremost; spillover benefits to industry occur in the form of more rapid development of selected technologies. In 1996, TRP partnerships were subsumed under the Dual-Use Applications Program.

5. SELECTED PUBLIC/PRIVATE PARTNERSHIP AREAS

Six selected public/private partnerships are overviewed in this section of the chapter. Each is discussed in an order reflecting their economic objective as set forth in the taxonomy of public/private partnerships in Table 1:

1) To leverage federal R&D (SBIR Program and CRADAs)

2) To enhance industrial competitiveness (ATP, NCRA research joint ventures, and SEMATECH)
3) To leverage industrial R&D (dual-use technology partnerships).

5.1. SBIR Program

The Small Business Innovation Development Act of 1982 authorized the SBIR program, and the program was then again reauthorized by the Small Business Research and Development Enhancement Act of 1992. Eleven federal agencies participate in the program. The five largest agencies being the Department of Defense, the National Aeronautics and Space Administration, the Department of Health and Human Services and its National Institute of Health, the Department of Energy, and the National Science Foundation.

The enabling legislation required each federal agency with an external R&D budget greater than $100M to set aside a stated percentage which was increased to 1.25 percent in 1986. When reauthorized in 1992, that percentage was increased to not less than 1.5 percent for the years 1993 and 1994, not less than 2.0 percent for the years 1995 and 1996, and not less than 2.5 percent for 1997 and subsequent years. SBIR awards are made in three phases. Phase I awards are designed to determine the technical merits of an idea. These awards last about 6 months to 1 year. Phase II awards, which last about 2 years, are to further develop the ideas in Phase I, if technically feasible. Phase III are more selective and generally involve follow-on contractual work with the funding agency.

With reference to the taxonomy of public/private partnerships in Table 1, the SBIR program's goal is to leverage the mission-oriented R&D of the financing federal agency. Obviously, each agency is supplying financial rather than research resources to the program. But, the SBIR program also provides infrastructural resources in the form of a unique and special funding environment. In particular, the program targets small- and medium-sized enterprises and minority enterprises for financial assistance.

There have been several survey-based studies related to the SBIR program. Based on information collected from samples of Phase II awardees, commercialization is taking place and it is having a measurable positive impact on society.[7]

5.2. CRADAs

Partnerships between industry and federal laboratories using a CRADA mechanism were made possible by the Federal Technology Transfer Act of 1986. While viewed by most as a technology transfer vehicle rather than a vehicle for direct funding of R&D, these mechanisms are conceptualized for the purposes of this chapter as an example of a public/private partnership that has an indented goal of leveraging the capabilities of federal R&D and thereby increasing the competitiveness of private R&D through transfers of knowledge.

In 1994, a preliminary evaluation of the benefits associated with cooperative R&D arrangements was conducted.[8] Based on detailed case studies of 10 representative CRADAs, the US Government Accounting Office (GAO) (1994) concluded that:

1) CRADAs offered opportunities for federal laboratories and industry to collaborate on research while still meeting their missions
2) Technology from federal laboratories was transferred to the private sector effectively, and this resulted in commercialized new products
3) Federal and industrial R&D programs were advanced in terms of scientific capabilities
4) Some CRADAs demonstrated a potential for long-term improvements in the economy, and overall health of the country

With reference to the taxonomy of public/private partnerships in Table 1, the goal of CRADA mechanisms is also to leverage the capabilities of federal R&D. Through CRADAs, the public sector provides financial resources in the form of in-kind labor, research resources in the form of access to specialized equipment in federal laboratories, and infrastructure resources by the very nature of the partnership mechanism.

To date, this GAO study represents the state-of-the-art. There has not yet been, except for the overview in Jankowski, Link, and Vonortas (2001), a critical evaluation of CRADA activity in terms of their effectiveness as a public/private vehicle for leveraging the capabilities of federal R&D.

5.3. Advanced Technology Program (ATP)

The ATP is a public/private partnership designed to enhance the competitiveness of industry. The enabling legislation, as discussed above, is explicit about that fact:

> The ATP ... will assist U.S. businesses to improve their competitive position and promote U.S. economic growth by accelerating the development of a variety of precompetitive generic technologies by means of grants and cooperative agreements.

Since its inception in 1990, as authorized by the Omnibus Trade and Competitiveness Act of 1988, ATP has generally held at least one open competition each year. While it is premature to judge definitively, the competitiveness consequences associated with ATP funding, there is initial evidence, based on case studies and survey information, that ATP has been successful in encouraging high-risk R&D projects that would not otherwise have been pursued; has helped U.S. industry compete in time-critical markets by accelerating their research agenda and thereby reducing time to market; has created new commercial opportunities for funded companies that are successful in their R&D efforts; and has promoted economic growth by creating an environment conducive for the permanent creation of selected new jobs.

With reference to the taxonomy of public/private partnerships in Table 1, the explicit goals of the ATP Program are to enhance competitiveness by underwriting selected research projects. Thus, by design, ATP represents a program for direct funding of private sector research through public sector financial resources. In addition, limited infrastructural resources are allocated to this program as well. ATP will, through its peer review of all proposals, marshal public resources to advise applicants about the scientific and technical merit of their research. This is a valuable public role because it affords companies that might not otherwise be able to obtain such advice, focused guidance even if their proposed project is not funded, and this guidance may facilitate a later re-application to the ATP or to other funding sources.

5.4. NCRA Research Joint Ventures

Unlike the ATP, research joint ventures (RJVs) formed as a result of the NCRA of 1984, have benefited more from federal infrastructural resources than federal financial resources. The environment created by NCRA and its 1993 amendment facilitated the formation of research joint ventures with assurances that any suspected antitrust violation will be judged on a case-by-case basis, and any related damage will be actual rather than treble. This is not to say that partners form research joint ventures with the intent of "hiding" purposeful antitrust violations, rather NCRA and NCRPA created an environment conducive to disclosure and therefore conducive to the formation of research partnerships.

Since 1982, the number of filed RJVs increased until about 1996 and then the number declined. There is some indication that the number is again increasing. While trends in NCRA research joint venture formation are not indicative of economic impacts so associated, the fact that the number of research partnerships, especially public/private (meaning that there is a federal organization as a member) partnerships, has increased over time is indicative of the fact that some benefits must be perceived by the public sector and by the private sector as a result of the federal organization being in the partnership. Theorists have modeled such benefits, as for example research cost savings, but to date no systematic empirical evaluations have been conducted on the benefits that members of such partnerships receive or on the spillover benefits to the related industry(ies).[9]

With reference to the taxonomy of public/private partnerships in Table 1, the 1984 NCRA and its 1993 amendment provided infrastructural resources in the form of a more conducive environment for the joint conduct of research.

5.5. SEMATECH

As discussed, with the passage of the NCRA in 1984, Congress explicitly recognized the importance of collaborative research relationships. The first, and perhaps one of the more notable research joint ventures that began shortly after the passage of the NCRA, was SEMATECH (SEmiconductor MAnufacturing TECHnology). Established in 1987 as a not-for-profit research consortium, its original mission was to provide a pilot manufacturing facility where member companies could improve their semiconductor manufacturing process technology. Its establishment came after the Defense Science Board recommended direct government subsidy to the industry in a 1986 report commissioned by the Department of Defense. It was thought that SEMATECH would be the U.S. semiconductor industry's response to the Japanese government's targeting of their semiconductor industry for global domination. Although the mission of SEMATECH has evolved over time, it generally has defined its mission around solving the technical challenges presented by sustaining a leadership position for the United States in the global semiconductor industry. Since its beginning, it has received Defense Department funding although funding ended at the end of fiscal year 1997 at the request of SEMATECH members.

SEMATECH sponsors three types of research projects. Joint Development Projects are part of a program in which a SEMATECH team of researchers partners with an external equipment supplier to develop a new tool, material, or process that supports phase requirements of future generation technology. Equipment Improvement Projects are designed to improve existing manufacturing equipment or systems from the perspective of competitive manufacturing. SEMATECH improvement projects reside at SEMATECH (rather than at a manufacturer's facility) and are short term and narrowly focused on production issues.

In contrast to NCRA research joint ventures in which the principle public resource brought forward to facilitate such partnerships is infrastructure in terms of creating an antitrust-friendly environment, SEMATECH represents what could be viewed in terms of the public/private partnership taxonomy in Table 1, as a "purer" example of direct financial resource support to achieve global competitiveness in a selected industry.

While there has not been a systematic evaluation of the impact of SEMATECH on the semiconductor industry as a whole, one focused investigation of the economic benefits members receive from their participation in the consortium was conducted by Link, Teece, and Finan (1996). They concluded from their detailed analysis of a sample of 11 representative and completed SEMATECH projects that companies benefited from sharing research results and that this sharing environment would not have existed without public funds to initiate the organization, and, in terms of member-specific benefits compared to member-specific costs, the ratio is approximately $3-to-$1.

5.6. Dual-Use Applications Program

In order to enable the Department of Defense to take full advantage of, and in many cases direct toward its military and defense needs, the research and technological capabilities of industry, Congress authorized the development of a direct funding mechanism called the dual-use technology partnership. After Congress approved the concept in 1990, and after the passage of the Defense Conversion, Reinvestment, and Transition Assistance Act of 1992, the Technology Reinvestment Project (TRP) began.

TRP was the Administration's so-called dual-use vehicle. It was an inter-agency program for grants to industry in pursuit of dual-use R&D and technology. Dual-use R&D and dual-use technology are those that are expected to be useful to both the military as well as to industry for commercial application.

TRP's initial strategy was to cost-share R&D projects in order to develop dual-use technology needed by both industry and the military. Department of Defense funding ensures that industry R&D can be directed toward the Department's special needs. Because this is technology that industry is expected to commercialize, private firms that receive TRP support encumber at least 50 percent of the R&D cost. The TRP strategy of dual-use partnerships leverages emerging industrial technology and embeds defense technology.

In 1996, TRP was subsumed under the Dual-Use Applications Program. Its mission was broader than TRP's; its mission is to prototype and demonstrate new approaches for leveraging commercial research, technology, products, and processes into military systems.

Dual-use technology partnerships, in terms of the taxonomy of public/private partnerships in Table 1, leveraged industrial R&D for the benefit of the military. Certainly public financial resources were utilized in this program since the military funds a minority portion of the R&D, but public infrastructure resources were also involved in that a specialized procurement environment (i.e., procurement process) was in place to identify and fund in an expeditious manner identified technologies without the bureaucracy of competitive bids.

A large number of focused studies have been conducted related to TRP. These studies almost exclusively have been case oriented in the sense that the objective of the study was to document that dual-use technology had in fact been developed.

The six public/private partnership areas are re-summarized in Table 2 in terms of the public/private partnership taxonomy in Table 1.

Table 2. Taxonomy of Public/Private Partnerships

	Economic Objective		
Aspect of Direct Funding by the Public Sector	Leverage Federal R&D	Enhance Industrial Competitiveness	Leverage Industrial R&D
Financial Resources	SBIR CRADAs	ATP SEMATECH	Dual-Use
Research Resources	CRADAs	—	—
Infrastructural Resources	SBIR CRADAs	NCRA RJVs SEMATECH	Dual-Use

6. PUBLIC ACCOUNTABILITY

Fundamental to the initiation of any public policy is recognition of accountability. With regards to technology-based public institutions, such as the six overviewed above, this accountability refers to being able to document and evaluate research performance using metrics that are meaningful to the institutions' stakeholders, meaning to the public.

Consider two different approaches to program evaluation. As Link and Scott (1998) have argued, the best evaluation technique for publicly-funded, publicly-performed research is based on a counterfactual method. In contrast, the best evaluation method for publicly-funded, privately-performed research consists of an analysis of spillovers.

Griliches (1958) and Mansfield, *et al.* (1977) pioneered the application of fundamental economic insight to the development of measurements of private and social rates of return to innovative investments. Streams of investment outlays through time—the costs—generate streams of economic surplus through time—the benefits. Once identified and measured, these streams of costs and benefits are used to calculate rates of return, benefit-to-cost ratios, or other related metrics.

In the Griliches/Mansfield model, the innovations evaluated are conceptualized as reducing the cost of producing a good sold in a competitive market at constant unit cost. For any period, there is a demand curve for the good, representing its marginal benefit to consumers, and a horizontal supply curve. Innovation lowers the unit cost of production, shifting downward the horizontal supply curve and thereby, at the new lower equilibrium price, resulting in greater consumer surplus. Additionally, the Griliches/Mansfield model accounts for producer surplus, measured as the difference between the price the producers receive and the actual marginal cost, summed over the output sold, minus any fixed costs. Social benefits are then the streams of new consumer and producer surpluses, while private benefits are the streams of producer surplus, not all of which are necessarily new because the surplus gained by one producer may be cannibalized from the pre-innovation surplus of another producer. Social and private costs will, in general, also be divergent.

The Griliches/Mansfield model for calculating economic social rates of return add the public and the private investments through time to determine social investment costs, and then the stream of new economic surplus generated from those investments is the benefit. Thus, the evaluation question that can be answered from such an evaluation analysis is: What is the social rate of return to the innovation, and how does that compare to the private rate of return? This is not the most appropriate question to ask from a public accountability perspective. The fact that the social rate of return is greater than the private rate of return may validate the role of government in innovation if the private sector would not have undertaken the research; but it ignores, for example, consideration of the cost effectiveness of the public sector undertaking the research as opposed to the private sector.

A different question should be considered when publicly-funded, publicly-performed investments are evaluated. Holding constant the very stream of economic surplus that the Griliches/Mansfield model seeks to measure, and making no attempt to measure that stream, one should ask the counterfactual question: What would the private sector have had to invest to achieve those benefits in the absence of the public sector's investments? The answer to this question gives the benefits of the public's investments—namely, the costs avoided by the private sector. With those benefits—obtained in practice through extensive interviews with administrators, federal research scientists, and those in the private sector who would have to duplicate the research in the absence of public performance—counterfactual rates of return and benefit-to-cost ratios can be calculated to answer the fundamental evaluation question: Are the public investments a more efficient way of generating the technology than private sector investments would have been? The answer to this question is more in line with the public accountability issues, and certainly is more in line with the thinking of public sector stakeholders, who may doubt the appropriateness of government's having a role in the innovation process in the first place.

There are important projects where economic performance can be improved with public funding of privately-performed research. Public funding is needed when socially valuable projects would not be undertaken without it. If the expected private rate of return from a research project falls short of the required rate called the hurdle rate, then the private sector firm will not invest in the project. Nonetheless, if the benefits of the research spill over to consumers and to firms other than the ones investing in the research, the social rate of return may exceed the appropriate hurdle rate. It would then be socially valuable to have the investments made, but since the private investor will not make them, the public sector should. By providing some public funding, thereby reducing the investment amount needed from the private firm or firms doing the research, the expected private rate of return can be increased above the hurdle rate. Thus, because of this subsidy, the private firm is willing to perform the research, which is socially desirable because much of its output spills over to other firms and sectors in the economy.

The question asked in the spillover method is one that facilitates an economic understanding of whether the public sector should be underwriting a portion of private-sector firms' research, namely: What proportion of the total profit stream generated by the private firm's R&D and innovation does the private firm expect to capture; and hence, what proportion is not appropriated but is instead captured by other firms that imitate the innovation or use knowledge generated by the R&D to produce competing products for the social good? The part of the stream of expected profits captured by the innovator is its private return, while the entire stream is the lower bound on the social rate of return. In essence, this method weighs the private return, estimated through extensive interviews with firms receiving public support about their expectations of future patterns of events and future abilities to appropriate R&D-based knowledge, against private investments. The social rate of return weights the social returns against the social investments.

The application of the spillovers model to the evaluation of public funding/private performance of research is appropriate since the output of the research is only partially appropriable by the private firm with the rest spilling over to society. The extent of the spillover of such knowledge with public good characteristics determines whether or not the public sector should fund or partially fund the research.

7. CONCLUSIONS

Herein, I have set forth a framework for conceptualizing technological alliances. This framework may be useful in that it identifies in a systematic way alternative alliances and their resulting technology elements.

As well, it may facilitate one's thinking of the paths through which public/private partnerships' technology-creating initiatives can work. Finally, whenever the public sector is involved purposefully in innovation, issues of public accountability arise and approaches to demonstrate accountability are set forth.

Chapter Three
COLLABORATIVE INNOVATION:
Rationale, Indicators and Significance

W. Edward Steinmueller
SPRU – Science and Technology Policy Research
University of Sussex

1. INTRODUCTION

Schumpeter (1947) argued that the ability to make innovation routine would be a major determinant of the future nature of capitalism. Schumpeter argued that the ability to make innovation a routine function in large companies would lead to a situation in which these companies would become indistinguishable in their scope of activity and size from other constituencies of society, and would therefore require some form of democratic control to resolve the diversity of social interests. Examination of the division of labor in innovation achieved through inter-organizational collaboration is therefore, ultimately, an enquiry into the future of capitalism (Pavitt and Steinmueller, 2002), a matter of some interest to those living in 'free market' economies.

Regardless of whether Schumpeter's vision is realized, the potential for collaborative innovation activities is of substantial importance in the foreseeable future, particularly in the advanced industrial economies where global sourcing of innovative activity has become another domain in which concerns about competitiveness may be raised. This issue is particularly important to those advanced societies, such as Canada, where augmenting and complementing the stock of domestic human resources supports skilled employment and the growth of innovative firms. Without such capacities to engage in a global division of labor in innovation, these economies will find it increasingly difficult to remain competitive against economies capable of mobilizing higher levels of inputs in generating and exploiting scientific and technological advance.

2. THE RESEARCH CONTEXT

In recent years, a new theory about the relation between knowledge generation and its use in the innovation process has emerged that stresses the systemic nature of innovation processes. The new perspective is concisely stated in the collaborative effort of OECD and Eurostat to define new methods of measuring innovation, the 'Oslo Manual':

> Systems approaches to innovation shift the focus of policy towards an emphasis on the interplays between institutions, looking at interactive processes both in the creation of knowledge and in its diffusion and application. (OECD/Eurostat, 1997)

In one sense, this new perspective is an extension of the early empirical literature on innovation such as Langrish, *et al.* (1972), which concluded that innovation processes were complex and interactive and, therefore, did not follow any of the prevailing innovation models in which events and actors were arranged in pre-determined 'linear' sequences such as basic research, applied research, and development.[1] In a second sense, the new theory is the consequence of investigations into the process of innovation within the firm related to the producer-user relationship in the innovative process (von Hippel, 1988; Lundvall, 1992) and to processes of supplier co-ordination (Asanuma, 1985). Detailed investigations of the characteristics of innovators and the processes of innovation have produced a large literature.[2]

These findings led a number of authors to suggest that there were distinct national (Freeman, 1982; Lundvall, 1992; Nelson, 1993) or sectoral (Edquist, 1997) systems of innovation. The new systemic theory of the innovation process has gradually gained ground against the still widely held view that innovation involves the progressive refinement of theory into practice – i.e. that innovation is applied science.

Systems of innovation approaches offer intriguing interpretive frameworks for explaining international differences in innovation performance. Despite their empirical origins, these theories are difficult to apply in a consistent way to the problems of international comparisons. It is difficult, for example, to isolate the effects of a particular national innovation system on industries in advanced industrialized economies that are increasingly linked to international sources of knowledge and experience. Even more difficult problems become apparent when it is recognized that these industries are engaged in comparative specialization *within* industrial sectors so that international trade in manufactures between advanced industrialized has a very strong *intra-industry* component. In effect, the national systems of innovation are producing distinct patterns of specialization within industrial sectors and it is therefore difficult to appraise the innovative performance of individual national systems.

Systems-oriented theories of innovation and the national innovation system approach provide an important conceptual background to the issues of networks, partnerships, and alliances. It would be simplistic, however, to exclusively attribute the growing attention to the structure and performance of collaborative arrangements to these developments in innovation theory.

As the OECD and Eurostat (1997) states, the new theories represent a convergence between industrial policy and science and technology policy. The growing concern with issues of national competitiveness that began in the mid-1980s led to concerns about the adequacy of support for innovation by national institutions including universities and national research laboratories as well as the innovative behavior of the private sector (Mowery, 1995). In addition, the European Union adopted specific policy objectives for building 'cohesion' among the member states with the view mitigating existing disparities in innovative performance and for strengthening the foundations for competitive performance. The European Commission sought to implement these policy objectives by imposing various requirements for receiving support in the Framework programs that were established to fund public and private research at the European level.[3]

Complementing these developments, empirical research beginning in the mid-1980s (Haklisch and Pouletty, 1986) indicated a rapid increase in the formation of technical alliances in the semiconductor industry. The unit of analysis, 'technological alliances,' as announced by press releases of companies, became a unit of analysis for empirical research (Hagedoorn and Schakenraad, 1990) and a subject for interpretation (Contractor and Lorange, 1988). During the 1990s, the efforts of national and regional authorities to promote collaborative arrangements provided further impetus to the empirical study of the formation and performance of collaborative arrangements. (Cohen, *et al.* 1997) An excellent summary of many of the theoretical issues and empirical approaches employed is Hagedoorn, Link and Vonortos (2000).

In short, the growing attention devoted to alliances, partnerships, and networks represents the confluence of a diverse collection of theoretical and empirical work on the structure and performance of innovation in advanced industrialized countries. Much of this work has been empirically motivated. The proliferation of inter-firm collaborative arrangements, and their association with attempts to maintain or increase competitive position in particular high technology industries such as semiconductors, raised practical questions about their effectiveness and influence. At the same time, scholars were attempting to deepen insights into the processes of industrial innovation by recognizing that knowledge creation often involved important processes of knowledge distribution and that such distribution often involved the formation of explicit inter-organizational linkages. Efforts to improve the competitive performance of national industries as well as to meet other policy objectives led policymakers to establish programs fostering linkages between industries and institutions such as universities and government research laboratories.

Evaluations of the effectiveness of such programs naturally followed and are a major academic research topic at this time.

3. RE-EXAMINING THE THEORETICAL FOUNDATIONS

Any effort to summarize the extensive literature on research networks, alliances, and partnerships that has emerged during the past two decades stands at risk of trivializing or, at best, over-simplifying the variety of theoretical foundations and traditions that this literature has employed. Nonetheless, it is important to concisely array the main arguments that have been employed in order to raise questions about whether the existing framework provides a sufficiently complete guide for further empirical investigation.

The central contention of this chapter is that the existing theory over-emphasizes a few specific mechanisms derived from the theory of the economics of information and division of labor at the expense of a richer and more complete behavioral theory of inter-organizational co-ordination and co-operation. The importance of this contention hinges upon the logical consistency of alternative motives and their empirical importance. Because of the absence of data to assess their empirical importance, this chapter focuses on three issues: 1) a critique of the existing approaches of inter-organizational collaboration in innovation processes, 2) a concise statement of alternative explanations and their potential empirical significance, and 3) the implications of the alternative theory for interpreting existing data and policy questions. The first two issues are dealt with in the following subsections (3.1. and 3.2) and the third is taken up in the two sub-sections of Section 4.

3.1. Collaborations as Division of Labor: Some Limitations in Current Approaches

Within the new 'systemic model of innovation', research collaborations are seen as a necessarily good thing since they improve the connectivity and tighten the coupling within an interactive system. Although some scholars (Katz and Martin, 1997) have expressed reservations about this argument and have argued for taking the costs of collaboration into account, these cautions have had very little discernible effect on policy formulation or empirical research. In effect there is a headlong rush to embrace the conclusions not only that more collaborations are better, but also that a standard for evaluation of the effectiveness of public support of scientific institutions should be the formation and extension of collaborative linkages.

The contention of this chapter is that it is useful to step back for a moment from this enthusiasm to examine the processes that it engenders or empowers. It is useful to simplify the argument by embracing the assumption that there is, indeed, a mutual interdependence between science and technology. Technology influences science, science influences technology. The two systems co-evolve. This leads to the requirement to examine more closely what consequences the technological system might produce for the conduct of science.

Firms make inter-organizational agreements in the context and for the purposes of achieving innovation.[4] The innovation process, however, is itself shaped by technological opportunities and by trajectories or tendencies in technological development. Innovation is a complex process that relates technological knowledge to market opportunity and the capabilities of the firm in exploiting both technological knowledge and market opportunity. The statement of the innovation problem in terms of the capabilities of individual firms raises an important question.

Why are inter-organizational agreements aimed at supporting or achieving innovation made and why does there appear to be an increase in their numbers over time? These are the classic defining questions of many of the studies that have been produced regarding networks, alliances, and collaborations. As in most areas where economists are active, the appearance of 'more' is associated with value. More agreements are made because their value is increasing. It is important to have some reservations about this conclusion – businesses are not immune from the impulses to follow fashion or trends. The tie between the value of agreements and their increasing frequency only serves to reinforce concerns about *why* such agreements are made.

The simplest explanation for collaboration is that it allows economizing of the research investment that is often necessary for innovation. In particular, collaboration allows the production of an information input that can be shared by the collaborators. In other words it is possible to achieve gains from employing the 'non-rivalrous' property of information as an economic commodity – the fact that multiple parties can use it without diminishing its amount. In addition, the processes involved in the production may benefit from a division of labor. For example, if the new information is to be found through a search process, strategies of dividing the area to be searched among multiple parties with an agreement that findings will be shared may produce an economic advantage for the participants. Finally, it is possible that the search processes may be pursued more effectively by using more expensive search methods. In all of these cases, collaboration provides a means for co-investment in a collective good. The benefits of the collaboration must be weighed against the costs of establishing the collaboration and the potential for diminished revenues if other members of the collaboration will compete in the markets for the new innovation, reducing the 'appropriability' of the benefits to be derived from the innovation.

A slightly more complex explanation for collaboration involves the recognition that firms may 'outsource' the production of inputs including the production of knowledge input to other organizations. Such outsourcing permits a division of labor in the innovation process and produces the classic principal-agent problem. The principal-agent problem is that the outsourcing firm must specify the expected outcomes from the research process and will have a strong motive to monitor the research process to avoid problems of 'moral hazard' in which the agent (the organization to which the task is outsourced) willfully fails to perform due to its pursuit of other interests or revenues. As in other principal-agent problems, the problem is to design a series of incentives that will avoid moral hazard problems and minimize the costs of monitoring. Shared equity arrangements such as research joint ventures are one organizational structure for creating compatible incentives. The difficulties of establishing a division of ownership or other rights to benefit from the results of such collaborations is a major problem given the uncertainties of the research process. For example, if one or more of the principals to a research joint venture undertakes a parallel stream of research to that conducted within the research joint venture, what value should be assigned to the research joint venture? Similar problems apply to the *ex post* division of returns within partnerships when the partners come to disagree about the relative levels of effort each of them has contributed to the outcome.

The models of division of labor in information search processes or of the functional or task specialization of the partners prevail in the current literature as the means for interpreting the growth of inter-organizational collaboration. A common explanation is that the amount of knowledge (usually scientific and technological information) needed for effective innovation is growing very rapidly and that collaboration provides a means for firms to avoid a dangerous upward spiral in the research costs necessary for innovation. Variants of this explanation involve the recognition that collaborators may be able to avoid duplicative research costs or enhance their collective market power over rivals by discovering information that will be more effective in the innovation process.

The prevalence of search and specialization explanations in the existing literature is the consequence of several factors. The first is the obvious attractiveness of the informational explanation to economists who are already influenced by the 'public good' argument for the public support of science.[5] The functional specialization argument is especially appealing to scholars of business studies who have interpreted outsourcing activities in terms of companies' increasing focus on 'core competencies.' Division of knowledge production activities is a natural extension of this general theory of outsourcing behavior. A number of scholars have also been attracted by arguments in favor of flexible specialization and the creation of production and distribution networks as an alternative to Fordist mass production models.

For these scholars, the formation of innovation networks is a complement to the further division of labor. Finally, the literature is influenced by the specific conditions of some of the industries in which alliances, networks, and collaborations are numerically prevalent. Several of these industries support the prevailing interpretation. For example, in pharmaceuticals the creation of a biotechnology sector focusing on production innovation serves as a function complement to the large pharmaceutical companies' capabilities in clinical testing and marketing. In semiconductors, the rapid escalation of the costs of pursuing the technological trajectory as well as the needs for innovations to meet environmental standards strongly suggested a division of innovative labor through collaborative agreements. Similar explanations have been produced for 'strategic' collaborations that involve combination of complementary capabilities.

The prevalence of search and specialization explanations in the literature are well-founded in economic and business logic. It is important to take these motives into account in explaining the formation and maintenance of collaborative arrangements. It is, however, possible that the division of labor has additional features whose implications have not been fully explored in explaining the nature or the number of collaborative arrangements. The contention developed in the next section that functional specialization involves a more complex division of labor in which co-ordination processes influence both the motives for forming collaborative arrangements and the activities conducted within these collaborations.

3.2. Collaboration as Co-ordination: New Explanations for Collaboration in a Modular World

Modern technology is heavily influenced by the effort to recognize further gains from the productive division of labor that Adam Smith noted in the case of the pin factory. His aphorism, 'the division of labor is limited by the extent of the market,' has been taken up in two highly influential papers Young (1928) and Stigler (1951). It is not surprising that the general trend towards global market liberalism we have experienced over the past decades has provided enormous impetus to the further division and specialization of labor. This context has had a profound influence on industrial structure, particularly in those industries based upon modern electronics or biotechnology that were established in the last half of the 20th century.

In modern technology, 'modularity' is an objective that increasing numbers of firms are pursuing in order to benefit from the specialized division of labor. The basis for modularity is the creation of effective specifications of the interfaces by which components can be assembled into systems. The design of modular systems produces a need for specific types of knowledge. For example, in designing a modern computer system component interfaces must have specific electronic properties in order to avoid bottlenecks in system processing speed. The components themselves carry with them further specifications that allow system integrators to evaluate the soundness of system designs with respect to heat dissipation or power consumption. At a first glance, most of this knowledge appears to be of a technological character in that it can be derived from some combination of existing scientific knowledge drawn from physics and practical experimentation. It is not obvious, however, that all of the problems that may be encountered in the design of component interfaces or their integration into systems can be addressed within the existing stock of technological and scientific knowledge. In this sense, the technological requirements of modularization and integration may generate an impetus for further knowledge generation. It is even less clear that these technological requirements will result in a smooth co-ordination across organizational boundaries, which is necessary for their effective implementation.

These observations provide the basis for creating a dichotomy between two types of knowledge generating activities. On the one hand, the generation of scientific and technological knowledge may involve search processes within domains that (as yet) are relatively unexplored or under-exploited. These processes may be referred to as involving a 'search' model of scientific and technological generation. It may involve explicit agreements to co-operate in collective search or it may involve the partitioning of knowledge generation activities related to innovation among participants (a division of innovative labor or functional specialization). On the other hand, the processes of modularization and integration involve somewhat different needs for the systems of knowledge generation. These processes comprise a 'co-ordination model' of knowledge generation.

In order to make the distinction operative it is necessary to argue that the goals and perhaps the organization of knowledge discovery processes differ in the contexts of 'co-ordination model' and 'search' models. Otherwise the 'co-ordination model' is simply a contextual labeling of the 'search model' or indistinguishable from the division of labor through functional specialization identified in the preceding section.

The argument that the knowledge search problem has different goals and may be organized differently is premised on the different requirements for the knowledge. In the case of the 'co-ordination' model the principal aim is to resolve ambiguities or uncertainties that may arise in modularizing systems and integrating components. In the language of modern economics, the problem is a constrained search for a satisficing solution. Solutions that have more general implications are ignored. For example, during the search for the electrical light bulb Edison discovered the Edison effect, wrote about in his laboratory notebook and got on with the search for an effective filament for the light bulb (the key interface in that particular system). Had Edison pursued the deeper implications of his discovery he could well have shared the Nobel laureate for physics won by J. J. Thompson for the discovery of the electron.[6] The point here is a simple one, the establishment of improved linkages aimed at improving co-ordination may lead to a failure to explore the more general implications of this knowledge. The assumption that *someone* will take up these more general implications is not necessarily well placed.

While this first point does provide some distinction between the search and co-ordination models, its consequences may not be large. Proceeding further, however, raises additional problems. Technological progress is a cumulative process. Building the next generation of systems involves heavy borrowing of the knowledge already accumulated in the building of the last generation. It is often the case that the new is distinguished from the old by a relatively limited progress along a trajectory governing the performance of a particular component.

Sometimes this advance is sufficient to introduce further problems in integration; often it is not. These observations suggest two implications for the 'co-ordination model' of demand for scientific knowledge. The first is that the search is likely to be path-dependent.

Increasing attention will be placed on generating knowledge in the vicinity of what is already being exploited. The value of knowledge in commercial activities is likely to be greatest when it can be directly exploited in solving the incremental problems of technological progress. Those increments are tied to past experience. In an important sense, the resulting knowledge search processes may suffer from diminishing returns from variety generation at the same time as they are offering greater commercial benefits. The other implication of this process is that components that are 'black boxed' and that are not on the path of cumulative improvement are not of very great interest. Hence, they do not direct any knowledge generation process. At this point it should be apparent that influence of the technological processes that produce the co-ordination process can have a substantial role in governing knowledge generation processes.

This is, however, not the end of the story. The division of labor made possible by modularization has specific implications of the location of economic activities. Increasingly components are produced at locations that are removed from the process of system design and integration. The consequence is that the production of knowledge is increasingly focused on interfaces or systemic integration problems rather than the components themselves. In an ideal world, these trends might create a productive division of scientific labor in which scientific knowledge generation activities followed components production. At present, however, this is not the case as components production is a highly competitive activity that is constrained by the risk aversion of the system integrator. To avoid creating new systemic integration problems, components are narrowly specified. This not only constrains the ability to, lends specific meaning to the subject of technological dependency, a topic for another day.

In developing the distinction between search and co-ordination directed knowledge generation, the previous remarks have been focusing on manufacturing and particularly the mass production of integrated systems. The same logic can be developed for a variety of other inter-organizational collaborations requiring the division of labor. The creation of medical systems such as blood analyzers or electro-cardiogram equipment involves similar divisions of labor and search processes. At an even more general level, the recent controversies in the pharmaceutical industry originating from the observation that tropical diseases are being ignored in the search for blockbuster products to treat the problems of aging can be understood as process of search constrained by the particular nature of the system to which the component (in this case a pharmaceutical) is expected to serve the needs of a modular system (in this case the health care system).

4. THE IMPLICATIONS OF CO-ORDINATION EXPLANATIONS FOR EXAMINATIONS OF INNOVATION NETWORKS, PARTNERSHIPS AND COLLABORATIONS

4.1. Empirical Research Implications

The distinction between search and co-ordination models of innovative knowledge generation have implications for interpreting existing data on research collaborations and for motivating new survey methods (i.e. involve answers that do not require a long lecture on how to answer). We begin by examining the existing sources of 'technological alliance' data such as those employed by Hagedoorn and Schakenraad (1992). This data is of considerable interest because it represents self-disclosure of arrangements between companies. It does, however, have some drawbacks including the lack of distinction between different motives for engaging in an 'alliance.'

The most likely distinguishing feature between search and co-ordination models of knowledge generation is the extent to which a collaborator has to take account of the action of another collaborative partner. However, this is not a sufficient condition as it is possible that collaborators engaged in a pure search process will also need to take account of the each others' actions. The relevant distinction is whether the collaborators actions are brought into mutual interdependence through the existence of a specific interface. Thus, the search collaboration can be distinguished from co-ordination collaboration by the intent of the co-ordination. On the one hand it may be described as avoiding duplication or benefiting from preliminary results in guiding further steps, indicating joint search. On the other hand, it may be described in terms that suggest that the results must be aligned, e.g.

we need to take into account the specifications we are working towards or the peculiarities of our collaborators' design. Such responses indicate the predominant activity involves co-ordination.

In industries subject to network externalities such as the information and communication technology industries it is also important to distinguish collaborations that are aimed at supporting the market success of a particular 'platform' in which the collaborators have a mutual interest. Identifications of the common technology platform or architecture that motivate the collaboration are sufficient to identify this form of collaboration. It is possible, however, that new platforms will require search as well as co-ordination activities and these may be distinguished by identifying whether the technological knowledge needed to implement the platform is within or outside the current state of art.

In general it is not possible to re-classify existing data sets of collaboration announcements because the intent of the collaboration is either unspecified or specified in very general terms. Research at the firm level needed to distinguish the search and co-ordination orientation of collaborations. As illustrated above, however, the information needed to make such a distinction is amenable to survey methods rather than requiring detailed case studies. The most significant problem is in devising means to weight the relative extent of both search and co-ordination activities when they exist in the same collaboration.

A second source of data on collaborative processes has been the identification of multiple inventors in patents and joint authorship in the scientific literature. In this case, even less is known about the extent of institutional partnership of the linkage (friends and former classmates may pursue their inventing or authoring interests regardless of their (current) institutional affiliation). Nonetheless, significant trends such as the identification of large numbers of institution-spanning co-authorships (Hicks and Katz, 1997) provoke important questions about the organization of scientific and technical research, suggesting to some (Gibbons, Limoges *et al.* 1994) that fundamental changes are underway in the locus of scientific knowledge generation. The possibility that some of these activities arise from the needs of firms for new knowledge to support co-ordination activities is significant for interpreting the effects of such activities on the research system. The existing literature is particularly concerned with the biomedical innovation where search activities may indeed be predominant. Nonetheless, identifying the role of co-ordination requirements as a source of collaborative research and innovation activities is worthy of future investigation. Medical practice involves a number of systemic issues and it would not be surprising that collaborative processes involving co-ordination were required to make effective innovations.

A third source of data is the membership of projects organized under the auspices of various funding bodies such as the European Commission who have specific objectives to encourage the formation of inter-institutional knowledge sharing. This data is generally more fragmented and situated, but provides some knowledge about how organizations form alliances when provided with specific incentives to co-operate with one another. Examinations of projects within the Telematics Application Programme of the 4th RTD&D Framework have indicated that universities have often been involved in 'search' processes and that many of the participants in such projects including industrial partners describe their motives for participating in terms of 'finding out' about new technological possibilities (ASSENT, 1999).

A fourth source of data is more sociological, attempting to identify 'invisible colleges' and similar collections of individuals that can be related to one another through patterns of citation, conference attendance, and committee membership. Research in this area has, so far, largely been focused on scientific communities. The principal opportunity for extending research of this sort to innovation networks is to examine the role of public and private standards organizations in the innovation process. Increasingly, these organizations are serving as a means for defining technological platforms or product architectures in which many participants have a direct interest. The collaborative activities necessary to achieve and maintain effective standards constitute a major source of 'technology alliances' or 'strategic partnerships.' Unfortunately, the substantial institutional and organizational complexity of such activities are unlikely to support survey based research.

Each of these data sources represents 'signals' about the very rich pattern of social exchange involved in innovation processes that stimulate technological and scientific research. At present, they provide substantial indications that changes are underway in collaborations involving technology. Unfortunately, it is not possible without further investigation to determine whether the increase in these activities is principally the consequence of the needs for further co-ordination as product architectures and production systems become more modular in order to take advantage of further opportunities for the division of labor. Neither is it warranted, however, to conclude that the proliferation of these activities as indicated by counts of collaborative arrangements is an indication the health of national or sectoral systems of innovation in discovering new technologies that will foster competitive advantage.

4.2. Policy Implications

Several policy implications follow from the re-interpretation of the existing data on collaborative arrangements as comprising a mixture (with unknown proportions) of search and co-ordination activities.

First, consider the aggregate statistics on all collaborative arrangements. If we take collaborations as a measure of the performance of the innovative system we can convince ourselves that public research laboratories and universities are 'disconnected' from or 'irrelevant' to knowledge generation. Recognizing the dichotomy between search and co-ordination research, however, suggests that a large share of collaborations are likely to be devoted to co-ordination research. For example, research of the type needed for resolving technological interface standards, involves problems of negotiation as well as discovery. It is not clear that public laboratories or universities have strong comparative advantage in these processes and it is not surprising that they are often not included. Incidentally, this does not mean that universities are entirely excluded from these processes since they may generate specific research needs that are undertaken by research contract rather than collaboration. The point here is that opening up the 'black box' of interactive knowledge generation reveals a much more complex clockworks than we are prepared to map or measure with our existing tools.

Second, a further feature of these clockworks describing the fine structure of interactive knowledge generation and distribution is that the number of collaborative agreements are inflated by the process of assembling 'tribes' of companies operating within the same product architecture. Companies are engaged in producing quasi-public goods involving harmonization of architectures and assembling of product platforms. Again, universities and public research laboratories can contribute to the exchange of information needed to articulate the knowledge resources of these tribes, but the knowledge generation processes may not be ones that open new avenues of scientific investigation.

The preceding two points indicate that universities and public research institutions may be structurally disadvantaged in the participation in knowledge co-ordination activities. Given the way that we count research collaborations they are therefore likely to under-participate in such arrangements.

The prevailing literature on the advantages of collaboration is very much focused on the benefits obtainable by rationalizing search processes. The claim that the costs of these search processes is what is driving the increase in collaborative arrangements may be a 'red herring.' If so, it becomes more important to assess disadvantages of collaboration such as its costs and its potential for diminishing variety in the search process.

In the UK as well a number of other countries, user participation has, increasingly, become either a desiderata or requirement of research funding. This is another instance in which the performance of science is being measured by the extent of collaboration. Such requirements foster a selection process in which university researchers are actively seeking out industrial and government partners. The costs of establishing and maintaining such relationships are significant for both universities and for private enterprises.

Two implications flow from these observations. First, the persistence of specific user partners creates incentives to build quasi-public or club goods rather than to pursue the open disclosure norm. This incentive is heightened by efforts to encourage universities to mind the intellectual property opportunities stemming from research. Second, the persistence of partnerships with specific users provides universities with the means to become more deeply involved in the co-ordination model of knowledge generation. The effects of these trends on the health of the knowledge generation process are uncertain.

In Europe, the 6th Framework programme is being planned based upon the existence of several different mechanisms for European funding support and co-ordination. Networks of excellence, large projects, and programs based on the participation of subsets of the EU members do not, inherently, favor either the search or co-ordination models of knowledge generation and distribution. The gross costs to participants in employing these mechanisms will, however, be large. In some knowledge domains, the scale of needed search activities will justify pursuing such mechanisms and they may prove effective, particularly if they succeed in maintaining variety generation in search processes. In other knowledge domains, the predominance of co-ordination requirements in the economic returns from knowledge generation and distribution will encourage the use of these mechanisms for supporting co-ordination rather than search. In this case, the aims of improving near term competitiveness and strengthening the research system may come into conflict.

5. SUMMARY

This chapter has identified a significant distinction between 'search' and 'co-ordination' activities in collaborative arrangements. The chapter contends that these have not been adequately considered in the prevailing literature on the innovation networks, collaborations and partnerships. The significance of this distinction is heightened by the opportunities provided by a growing international division of labor in knowledge generation as well as productive activities. To the extent that collaborations are principally engaged and effective in solving the relatively 'local' problems associated with the effective construction of modular systems there will be ample motives for the continued formation of collaborative arrangements. Associating the growth in such collaborations with the dispersion of traditional research activities associated with search or the division of innovative labor may, however, be dangerous. It is, in principle, important to distinguish the purpose of collaborations before using them as a measure of innovative performance. In particular, it is hazardous to conclude that the relative paucity of university and public research laboratory participations in such collaborations is an indicator of these institutions 'disconnection' from knowledge generation or distribution. As in many areas of social science, the interpretation of data within a particular frame of reference may lead to incorrect conclusions if that frame of reference proves to be limited.

PART II

MEASUREMENT

Chapter Four
AN ANALYSIS OF PATTERNS OF COLLABORATION IN CANADIAN MANUFACTURING AND BIOTECHNOLOGY FIRMS

Frances Anderson, Chuck McNiven and Antoine Rose
Science, Innovation and Electronic Information Division, Statistics Canada

1. INTRODUCTION

Innovation plays a pivotal role in economic development. The processes of innovation at firm and national levels have been the subject of much investigation and analysts have sought to clearly understand the innovation process. One very important aspect of the innovative process is how and why firms use alliances, collaborations and networks with external partners in the development of their innovative products and processes.

The two surveys analyzed in this chapter (Survey of Innovation 1999 and, Biotechnology Use and Development Survey 1999) probed the firms on their collaborative behavior by gathering information on types of partners and purposes for entering into cooperative/collaborative arrangements. This information can in turn be compared to other characteristics of the firms such as size, industry, R&D performance or stage of development.

There is an abundant literature on collaboration and alliances. First, Bressand and Kalypso (1989) provided an economic justification for networks of alliances by defining them as a "closed set of selected and explicit linkages with preferential partners in a firm's space of complementary assets and market relationships, having as a major goal the reduction of static and dynamic uncertainty..." where alliance is characterized by complementarity and the need to distribute the uncertainties amongst the partners.

Focusing on innovative networks, Freeman (1991) noted changing patterns of collaboration in the 1980s and abundant evidence of a strong upsurge of various forms of research collaboration, especially in the new generic technologies. There was also ample evidence of a qualitative change in the nature of relationships involving more research associations and government R&D projects and programs. In their book on *The New*

Production of Knowledge, Gibbons *et al.* (1994) emphasized the fact that knowledge is increasingly produced in an application context and is transdisciplinary in nature. In the firm context, this is the recognition that the knowledge required to develop new applications (innovations) cannot be entirely possessed or developed within the firm.

Exploring the notion of uncertainties using transaction cost analysis, DeBresson and Armesse (1991) developed a two dimensional space evolving between standardized and stable technology and technological uncertainty on one axis and between clear property rights and appropriation uncertainty on the second axis. They argued that innovation networks would be observed in areas where firms face both technological and appropriation uncertainties. Combining the notion of complementarity and the need to redistribute the uncertainties, Hagedoorn (1993) proposed an overview of motives for strategic (technology) alliances:

1) Sharing and further advancement of R&D (because of increased complexity and intersectoral nature of new technologies)
2) Capture some of the capabilities, knowledge or technologies of the partner through joint innovation projects.
3) Market access and technology development

Bayona *et al.* (2001) added market access and the reduction and sharing of uncertainties and costs to this list.

The term collaborative/co-operative arrangement or alliance describes a set of business processes involving some form of defined partnering between enterprises or businesses. In generic terms, alliances are formal or informal arrangements between two or more organizations to achieve specific goals. Hagedoorn (1990) defines alliances as "inter-firm agreements that can reasonably be assumed to affect the long-term product market positioning of at least one partner" (p. 20). Niosi (1996) suggests that two prime methods exist for firms to enter into alliances; memorandums of understanding (MOUs) linking firms for a project bound by specified parameters and joint ventures. Alliances range from permanent agreements to carry on business to formal financial agreements and exchanges, to informal exchange arrangements of knowledge, technology or equipment sharing. The delimiting parameters of the relationship could include time, money, knowledge, product development, market penetration, and geographic characteristics, singly or in combination, making the strategic alliance a flexible business relationship particularly suitable for enterprises with different capabilities.

We also observe that networks of alliances are made of firms of different sizes and various stages of development. Alliances between incumbents and emerging firms play a role in new technology adoption and the subsequent transformation of an industry. "In biotechnology, we are witnessing a transformation of the traditional, chemical-based pharmaceutical industry into the newly emerging biopharmaceutical industry... This transformation is mainly the result of extensive interfirm cooperation between incumbents and new entrants". (Rothaermel, 2001, p. 1245)

Building on this body of literature, this chapter addresses the following questions:

1) Do patterns of innovation collaboration vary by firm size? For example, are small firms more likely to use collaborative partners which are part of their supply chain (suppliers or clients)?

2) Do patterns of collaboration vary by the extent of internal commitments to R&D? For example, do firms which undertake R&D in their own R&D departments have different patterns of collaboration than firms that rely on outsourcing R&D?

3) Do patterns of collaboration vary by the industry within which the innovative firms are located? For example, are patterns of innovation collaboration in the pharmaceutical industry significantly different that what is found in the clothing industry?

4) Do patterns of collaborations vary according to the stage of product development of the firm? For example, are firms that are in the research and development stage of product development more involved in partnering with universities than are those firms that are involved in the regulatory stage of product development?

5) What is the relation between collaboration and the raising of capital? For example, are firms engaged in collaborations more successful in raising capital?

The first section will present the findings from the Survey of Innovation 1999 and the section second section will present data from the Biotechnology Use and Development Survey 1999. The conclusion discusses the issues that have arisen from the empirical analysis of the two surveys.

2. METHODOLOGY AND DEFINITION OF TERMS

Data from two Statistics Canada surveys, the Survey of Innovation 1999, which surveyed the Canadian manufacturing sector, and the Biotechnology Use & Development Survey 1999, which surveyed biotechnology in manufacturing and other sectors of the economy, will be the basis of this chapter's examination of patterns of collaborative arrangements by innovative firms.

The methodology in the two surveys differs in respect of the size of the firm, the statistical unit of observation and the specific questions that have been asked.[1] Care needs to be taken in the interpretation of the results and references.

3. PART 1, SURVEY OF INNOVATION 1999

The Survey of Innovation 1999 surveyed 5,800 manufacturing firms in Canada. The survey uses the Oslo Manual guidelines for measuring innovation (OECD/Eurostat, 1997). Innovators are defined as those firms that have introduced new or significantly improved products or processes during the previous three years. The survey found that 80% of manufacturing firms in Canada were innovative. Furthermore, of these innovative firms, 33% were involved in cooperative and collaborative arrangements.

Firms involved in cooperative and collaborative arrangements were asked to identify their partners. The Survey of Innovation 1999 does not provide data to respond to the question on raising capital. A discussion of the four other questions posed in the introduction is found below.

In the Survey of Innovation 1999 firms were asked the following question in respect of innovation collaboration.

> Was your firm involved in cooperative and collaborative arrangements with other firms or organizations to develop new or significantly improved products (goods or services) or production/manufacturing processes during the last three years, 1997 to 1999?
>
> The definition provided to the firms was as follows:
>
> Cooperative and collaborative arrangements involve the active participation in joint projects between your firm and other firms or organizations in order to develop new or significantly improved products (goods or services) and/or production/manufacturing processes. Pure contracting-out work, where there is no active participation, is not regarded as collaboration or cooperation.

Firms that they are involved in cooperative and collaborative arrangements were further asked to identify their partners. This data is the focus of the following discussion.

3.1. Size[2]

In very general terms, firms of all sizes collaborate more with private sector partners (suppliers, clients, consulting firms and competitors) than with public sector partners (university and federal and provincial research institutes). (Table 1)

Table 1: Collaboration Partners of Innovative Manufacturing Firms, By Size, 1997-1999

	Percentage of Innovative Collaborators Collaborating with Partner			
	Small	Medium	Large	Total
Suppliers	70%	69%	78%	71%
Clients	65%	67%	62%	65%
Consulting firms	34%	46%	45%	39%
Competitors	34%	38%	35%	35%
Universities	21%	26%	29%	24%
Firms in corporate group	17%	23%	36%	23%
Federal research institutes	20%	21%	22%	21%
Provincial research institutes	16%	17%	15%	16%

Source: Statistics Canada, Survey of Innovation 1999

The survey found as well that different size firms have different patterns of collaboration when the different types of partners are examined. The larger the firms are, the more likely they are to collaborate with their suppliers, whereas small and medium sized firms are more likely to collaborate with their clients. Larger firms are also more likely than smaller ones to collaborate with other firms in their corporate group. The higher rates of collaboration may reflect the fact that larger firms are more likely to be part of a corporate group than are the smaller ones.

As for collaborating with public sector partners, the larger the firm the more likely it is to collaborate with universities, whereas the percentage of firms of all sizes that collaborate with federal and provincial research institutes are relatively similar.

The results of the survey indicate that perhaps some important types of collaborators are missing from the list that was provided to respondents. Firms were asked to indicate other types of partners that were not on the list. Some of the "missing" partners indicated by the respondents were industrial research associations, private R&D companies and joint ventures.

3.2. Internal Commitment to R&D

Firms were asked if they carried out R&D in a separate and distinct R&D department. A positive response to this question can be taken as an internal commitment to R&D.

Findings presented in Figure 1 show that there are significant differences between the collaboration patterns of firms with in-house R&D and those without. Firms with in-house R&D are significantly more likely to collaborate with public sector organizations (universities and provincial and federal research institutes). They are more likely to collaborate also other firms in their corporate group and consulting groups.

In contrast, there is relatively little different between the percentage of firm with in-house R&D and those without when it comes to marketplace partners (competitors, suppliers and clients).

Figure 1: A Comparison of the Percentage of Collaborating Innovative Manufacturing Firms With In-house R&D and Those with No In-house R&D, by Type of Partner, 1997-1999

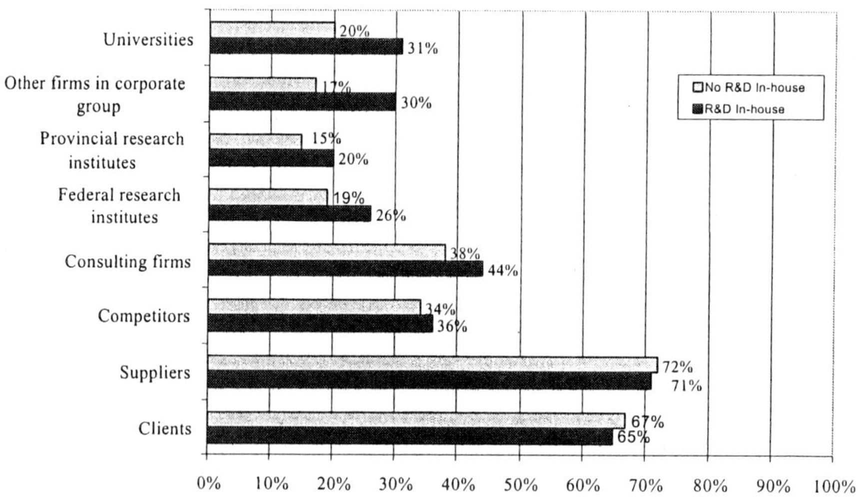

Source : Statistics Canada, Survey of Innovation 1999

3.3. Stage of Development

Although the Survey of Innovation did not explicitly address the issue of stage of development, some of the questions asked in a section on "reasons for collaborating" can be taken as proxies for stages of development.

For the research phase (accessing R&D) larger firms are more involved in collaboration for this purpose than are smaller firms. In the development phase (prototype development and scaling-up), large and medium size firms are equally involved in collaboration for this purpose and small slightly less. In the commercialization phase (accessing new market and accessing new distribution channels), small firms are particularly involved in collaboration for this purpose.

Figure 2. Reasons for Collaboration

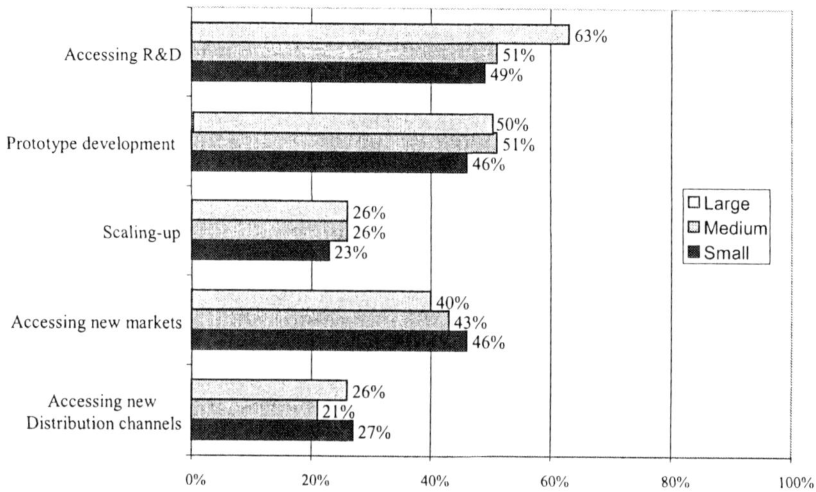

Source: Statistics Canada, Survey of Innovation 1999

3.4. Industry of the Firm

Thirty-three percent (33%) of all innovative manufacturing firms were involved in collaborations. The percentage of firms in different industries varies considerably from a high in pharmaceuticals where 61% of innovative firms were involved in collaboration for innovation to a low of 16% for innovative firms in clothing manufacturing. (Table 2)

Table 2: Percentage of Innovative Manufacturing Firms Involved in Innovation Collaboration, by Industry, 1997-1999

Industry[3]	%	**Industry**	%
Pharmaceuticals	61%	Motor Vehicle	38%
Semiconductor/Electronic Equipment	57%	Textile Products	36%
Computer Equipment	53%	Leather	36%
Aerospace	51%	Miscellaneous Manufacturing	34%
Instrumentation/Magnetic-Optical Media	50%	Petroleum/Coal	34%
Beverage/Tobacco	50%	Plastics and Rubber	33%
Communications Equipment	47%	Non-Metallic Mineral	32%
Electrical Equipment/Appliances	44%	Food	32%
Railway/Ship	42%	Textile Mills	32%
Plywood/Engineered Wood	40%	Fabricated Metal Products	30%
Agriculture/Construction/ Mining and Industrial Machinery	40%	Printing	27%
Chemical	39%	Sawmills/Wood Preservation	22%
Paper	39%	Furniture	20%
Primary Metal Manufacturing	39%	Other Wood Product	18%
Other Machinery	39%	Clothing	16%

Source: Statistics Canada, Survey of Innovation 1999

An interesting question, of course, is why there are such different patterns of collaboration in the different industries, not only in terms of the percentage of firms in an industry that collaborate (Table 2) but also in terms of the partners with whom they tend to collaborate (Table 3). Martin and Scott (2000)[4] have developed a taxonomy of "modes of innovation". Although their purpose was to characterize the sources of sectoral innovation failure and the appropriate policy instruments to address these failures, the taxonomy will be used in this chapter to provide a framework to discuss collaborative relations. Industries are classified into the following four categories:

Type 1: Developing of inputs for using industries: These are the intermediate good industries where the "predominant form of innovation is the development of higher quality products that will be used as inputs in vertically integrated industries" (p. 440). Examples are equipment-producing and software industries.

Type 2: Application of inputs developed in supplying industries: These are the down-stream industries that innovate "by adapting products and processes developed in upstream industries to their own commercial needs". Examples are agriculture and light industry.

Type 3: Development of complex systems: These industries produce complex systems where the cost of R&D projects is very high and there is great risk for those firms that do not stay on the technological frontier. Because of the steep learning curve, firms that do not keep up risk extinction. Innovation in these industries involves common standards for infrastructure technologies. Examples are aerospace, electrical and electronics technology and telecom/computers.

Type 4: Applications of high-science-content technology: Innovation in these industries relies on a technology base with a high science content. Examples are industries involved in biotechnology, chemistry, materials and pharmaceuticals.

Table 3: Top 5 Collaborating Industries, by Type of Partner, by Percentage of Innovators that Collaborate, 1997-1999

University	**Federal Government Research Institutes**
1. Semiconductor Industry (68%)	1. Instruments and Magnetic-Optical Media (44%)
2. Pharmaceutical (46%)	2. Veneer/Engineered Wood Products (37%)
3. Electrical Equipment (37%)	3. Primary Metal Products (35%)
4. Petroleum/Coal (36%)	4. Chemical (30%)
5. Chemical (36%)	5. Electrical Equipment (29%)

Clients	**Suppliers**
1. Semiconductor Equipment (88%)	1. Computer Equipment (96%)
2. Instruments and Magnetic-Optical Media (81%)	2. Leather Products (95%)
3. Computer Equipment (77%)	3. Petroleum/Coal Products (93%)
4. Communications Equipment (77%)	4. Beverage/Tobacco Products (92%)
5. Clothing (76%)	5. Furniture (89%)

Competitors	Other Firms in Corporate Group
1. Aerospace (61%)	1. Semiconductor Equipment and other Electronic Components(47%)
2. Pharmaceuticals (58%)	2. Beverage (43%)
3. Railway/Ship (54%)	3. Petroleum/Coal (43%)
4. Beverage (52%)	4. Pharmaceutical (43%)
5. Petroleum/Coal (50%)	5. Food (36%)

Source: Statistics Canada, Survey of Innovation, 1999

To what extent is the above taxonomy of "modes of innovation" useful in explaining the observed collaboration patterns. It is striking to note that of the 15 industries with the lowest percentage of collaborating innovative firms (Table 2), 12 of these industries are those that apply the inputs developed in supplying industries (Type 2). The relative lack of collaboration can be explained by the fact that the principal mode of innovation is through buying and adapting input generated by other industries. Only two of the industries that apply the inputs developed in the supplying industries are found in the top-15 industries (Plywood, Veneer and Engineered Wood and Primary Metal Manufacturing). Both of these industries have a high percentage of innovative firms collaborating with federal government laboratories, indicating perhaps special R&D programs directed towards these natural resource based industries.

In the top 15 industries (excluding the two mentioned above), in terms of percentage of collaborating innovative firms, one finds a mix of the three types of industries (Types 1, 3 and4). In terms of firms with the highest percentage of collaborations (top-5) with specific partners, there is some differentiation. Science based industries (Type 4) and Equipment manufacturers (Type 1) are the highest in terms of collaboration with universities as they are with federal laboratories (with the exception of the two Type 2 industries discussed above). Both of these types appear to have a requirement for access to R&D in public institutions.

Equipment manufacturers (Type 1) (with the exception of Clothing) have the highest percentage of innovation collaboration with clients which is an expected result as they are providing intermediary goods to other industries and collaboration with clients can be considered as critical to the development of their new products.

In terms of cooperation with competitors, two complex products industries (Aerospace and Railway/Ship Building) were very high. Collaboration with competitors for these industries is an expected result as the cost of the development projects and the need of standards for infrastructure technologies would require inter-firm collaboration. It is not so clear why there is particularly high collaboration with competitors in the Pharmaceuticals, Beverage and Petroleum/Coal.

In general terms, one can conclude with this preliminary analysis of collaboration and the taxonomy of "modes of innovation", that collaboration patterns do appear to be related to the mode of innovation.

4. PART 2: BIOTECHNOLOGY

The Biotechnology Survey was a technology-focused survey. It surveyed a diverse set of activities – emerging and evolving technologies in contrast to the Innovation survey's focus on a range of industries categorized by a well-understood classification system. Both types of survey provide insight into innovation and collaborative arrangements and each contributes to our understanding of collaborative arrangements and innovation, by giving a different perspective on the subjects. This section of the chapter provides a perspective based on the Biotechnology Use and Development Survey 1999. Selected characteristics on biotechnology firms appear in Table 4.

Table 4. Selected Biotechnology-Related Characteristics of Biotechnology Firms by Firm Size (1999)

	Revenue ($000,000)	R&D ($000,000)	Collaborative Arrangements	Biotechnology Employees	Number of Firms
Small (50 or fewer employees)	249	256	411	2,935	270
Medium (51-150 employees)	295	106	123	1,343	51
Large (151 or more employees)	1,404	465	160	3,470	37
Total	1,948	827	694	7,748	358

Source: Statistics Canada, Biotechnology Use and Development Survey 1999.

The data provided in this table focus on biotechnology innovators, firms that are using biotechnology to develop new products and processes. To become a biotechnology innovator, the respondent must meet three criteria: the respondent is using at least one of a list of biotechnologies; R&D activities in biotechnology are performed and; the respondent is currently developing products or processes that require the use of biotechnology.

The respondents to the Biotechnology Use and Development Survey were asked the following questions on cooperative/collaborative arrangements:

Was your firm involved in biotechnology-related cooperative/collaborative arrangements with other companies or organizations in 1999?

The definition provided to the firm was as follows:

Cooperative and collaborative arrangements involve the active participation in projects by your company and other organizations in order in order to develop and/or continue work on new and significantly improved biotechnology processes, products and /or services. Pure contracting/out is not regarded as collaboration.

These questions and definitions are similar to the ones asked of the Survey of Innovation 1999 respondents. The question from the Biotechnology Use and Development Survey is not directly related to innovation per se, but that is clearly mentioned in the definition. The two same key elements are present: a sustained participation in projects that goes beyond pure contracting-out and, this collaboration is related to the development of new or significantly improved products and processes. An additional restriction found in the Biotechnology Use and Development Survey is the focus on biotechnology innovations.

Biotechnology is an enabling technology. It is an engine for knowledge generation and the development of new products and processes. During the development of products or processes involving biotechnology, firms face multiple challenges. Each of these challenges requires specific competencies/capacities that may or may not exist within the firm. Feldman (2001) states "...biotech is a very knowledge-intensive industry and progress requires complementary assets that reside in different types of organizations." (p. 1) One option is to form alliances with others. Established firms that want to adopt biotechnology must obtain and develop knowledge about biotechnology. Start-up and spin-off firms may seek a different set of competencies. These are often related to the commercialization process. Alliances are a means for both to obtain new competencies. Niosi (2000) found that "Alliances are key for emerging firms that need enormous resources – knowledge, facilities, skilled personnel and capital. It is no surprise then that three-quarters of the companies (47/60, or 78 per cent) conducted alliances" (p.16)).

4.1. Size[5]

Based on evidence from the survey; 62% of the 358 biotechnology firms reported a total of 694[6] co-operative arrangements. Small firms accounted or 59% of those arrangements, with 168 of the 270 small firms entering arrangements, followed by 28 of the 37 large firms with 23% of the arrangements. Figure 3 demonstrates the average number of collaborative arrangements per firm based on firm size.

Figure 3. Average Number of Collaborative Arrangements by Firm Size Among Firms that Collaborate

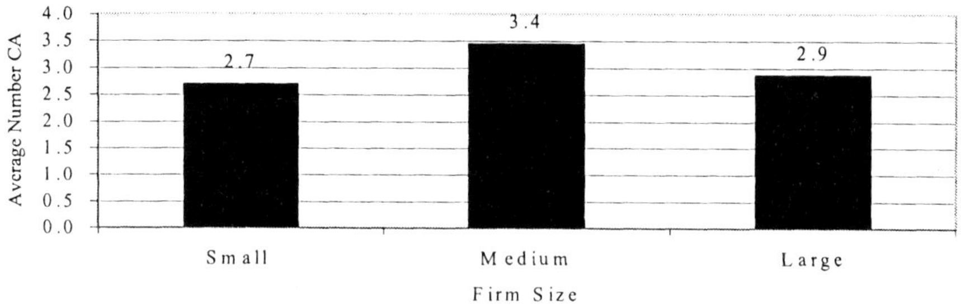

Source: Statistics Canada, Biotechnology Use and Development Survey 1999

Lerner and Merges (1998) found that in the United States large pharmaceutical or medical firms ally with small biotech firms lacking complementary assets such as sales forces and manufacturing, but which are rich in innovative ideas or research. Larger firms benefit from smaller firm's innovation speed and freedom, and the smaller firms benefit from the larger firm's infrastructure. Lerner and Merges (1997) asserts young firms "lack the financial resources to effectively introduce a new product" and "lack complementary assets (i.e. sales forces and manufacturing know-how). As a result, strategic alliances between small, research-intensive firms and larger corporations have become common-place" (p1). Results from the survey show that small firms allied with large firms 31% of the time. Alliances are not only established between firms. Niosi (2000) found that two-thirds of firms conducting alliances had university partners, mainly aimed at basic research (p.16).

4.2. Internal Commitment to R&D

Using the data from the Biotech survey, a biotechnology firm is defined as a firm that use biotechnology, performs biotechnology R&D activities and, are developing products and processes that require the use of biotechnology. Therefore, all the firm characteristics reported here are related to firms engaged in R&D activities. We cannot report of differences between firms engaged, or not engaged, in R&D as it is the case with the Survey of Innovation 1999.

4.3. Type of Partners

Participants in alliances come from business, academia and government sectors in almost any combination. Relationships can be vertical, between vendor and customer, or horizontal, between vendors, local or global and occur between competitors. Firms reached 194 agreements with universities/hospitals, 107 agreements with government departments/agencies and 336 collaborative arrangements with other business both smaller and larger. Overall firms joined with universities/hospitals in 30% of arrangements, followed by large firms at 28% of alliances. The majority of small firms indicated an equal number of arrangements with larger firms or universities/hospitals - 31% of the arrangements. Smaller or same sized firms made up 24% of alliances and government department/agencies 14%. Table 5 provides detailed information on partnership type.

Table 5: Total Number of Cooperative/Collaborative Agreements by Partner Type and Firm Size 1999

	Firm Smaller or Equal Size	Larger Firm	Government, Department or Agency	University/ Hospital	Total
Small	98	128	57	128	411
Medium	..	..	27	38	141
Large	..	..	23	28	86
Total	156	180	107	194	638

.. Figures not available

Source: Statistics Canada, Biotechnology Use and Development Survey, 1999

Alliances serve as a method for larger firms to secure rights to technology by establishing market, research or manufacturing relationships with smaller firms. The smaller firm benefits from access to cash, business expertise, complementary technologies, or the ability to combine marketing, distribution, manufacturing or financing opportunities. Not to be overlooked is the credibility given to a smaller firm when allied with a larger firm. This view is echoed by Senker and Sharp[7] (1997) who developed the idea that the smaller dedicated biotechnology firm requires a relationship with a larger firm in order to attract venture capital funds. Larger firms report a preference for alliances with smaller firms and smaller firms reporting more alliances with larger firms.

4.4. Purpose of Collaboration by Size

The purpose of collaborative arrangements varies. Firms may seek an alliance for R&D as a method to reduce risk associated with efforts to solve problems or to further advance innovation. Canadian biotech firms entered into collaborative arrangements with the purpose of research and development 33% of the time, the number one reason firms entered collaborative arrangements. Alliances often occur in young industries where speed and flexibility are key components to survival. This may be especially true in the biotechnology industry where firms race to patent innovations (Baum and Silverman, 1998). In Canada 9% of the arrangements were sought to protect intellectual property and 6% for regulatory affairs. Firms also conducted alliances for other R&D purposes such as accessing knowledge (22%) and prototype development (18%). Table 6 highlights the purpose of collaborative arrangements based on firm size. The 223 biotechnology firms with collaborative arrangements reported 694 collaborative arrangements, citing 591 different purposes, demonstrating the diversity of collaborative arrangements.

Table 6: Purpose of Collaborative Arrangements by Firm Size, 1999

Firm Size	Conduct R&D	Regulatory Affairs	Access Knowledge	Prototype Development	Access Markets	Access Capital	Intellectual Property Protection	Other	Total
Small	134	27	81	67	37	..	35	..	403
Medium	32	..	31	..	..	..	..	..	125*
Large	30	..	15	..	..	..	..	..	64
Total	195	33	128	104	48	..	51	..	591

.. Figures not available
* Use with caution, unreliable due to high coefficient of variation

Source: Statistics Canada, Biotechnology Use and Development Survey, 1999

Examining the survey data using the firm size category and purpose of collaborative arrangements demonstrates differences in purposes based on firm size. Small firms report all reasons to enter collaborative arrangements, but larger firms are more selective in their purpose, in fact citing three main reasons to collaborate: prototype development; accessing knowledge; and, R&D purposes. Medium sized firms opt for prototype development and accessing knowledge. The largest firms are especially evident as collaborators for R&D purposes.

Figure 4. Purpose of Collaborative Arrangements by Firm Size

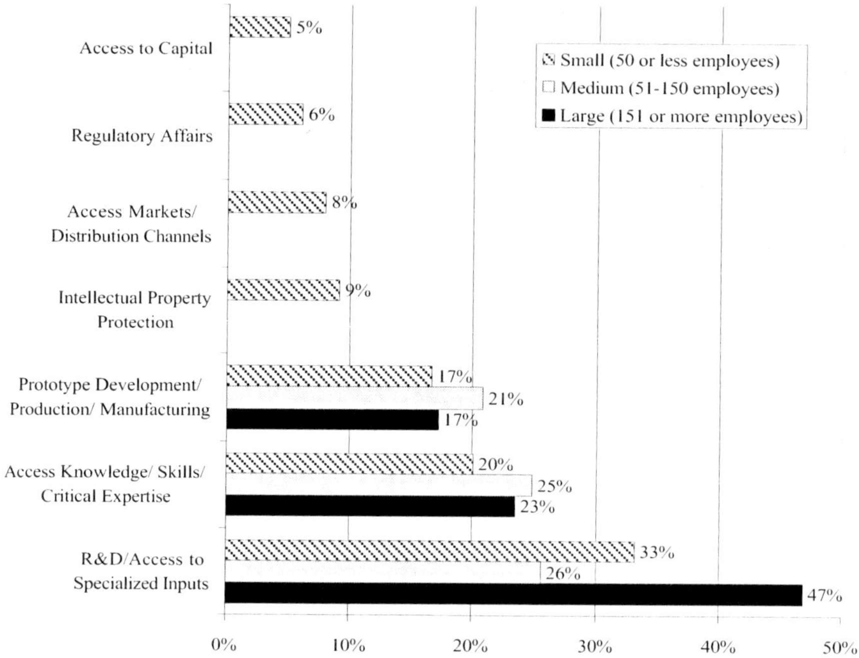

Source: Statistics Canada, Biotechnology Use and Development Survey 1999

Alliances not only serve the traditional practices and needs of business but also serve as the channels through which information and knowledge are transferred. The alliances form part of the network of relationships. Hagedoorn (1993) states that biotechnology alliances simultaneously accomplish multiple purposes. The type of partner or style of partnership will vary with the particular challenge or challenges faced. For example in the early stages of product development firms seek more technical help through alliances with universities or research facilities. At later stages of development gaining regulatory approval requires extensive specialized knowledge, and at the commercialization stage firms may be looking for distribution channels.

4.5. Purpose of Collaboration by Stage of Development

Further evidence suggests that firms enter into collaborative arrangements for different purposes based in part on their stage of development. Firms are more likely to enter collaborative arrangements for R&D purposes when in the R&D stage than the clinical and market stages, although these do remain as prominent reasons for collaborative arrangements. Prototype development increases as the stage of development evolves, as do regulatory affairs and to a lesser extent access to capital. Small biotechnology firms entered into collaborative arrangements to conduct R&D (33% of arrangements for small firms), access knowledge (20%) and prototype development (17%). The following section of the chapter discusses the relationship between collaborative arrangements and raising capital.

Figure 5: Purpose of Collaboration by Stage of Development

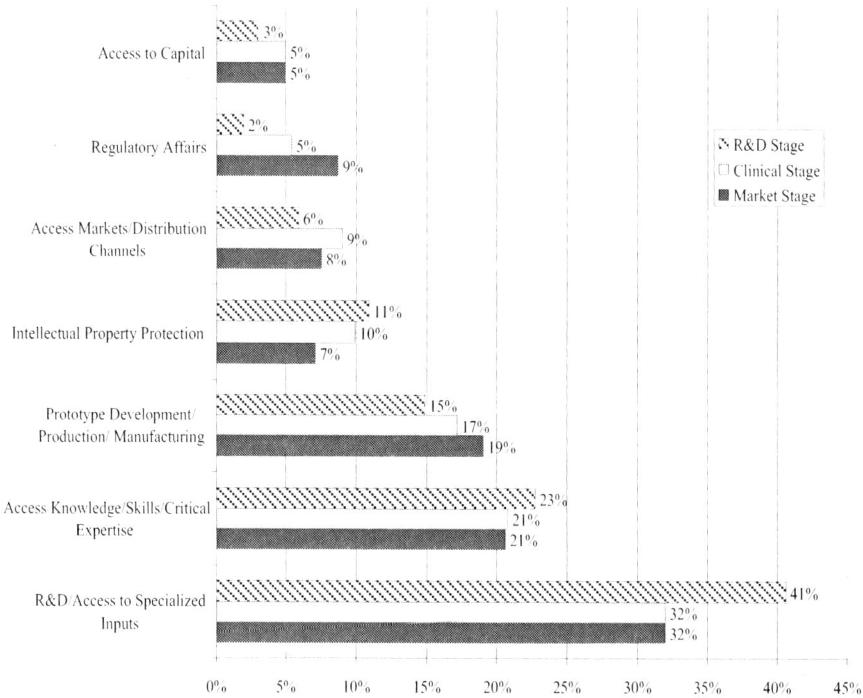

Source: Statistics Canada, Biotechnology Use and Development Survey, 1999

4.6. Mode of Innovation

Biotechnology firms are mostly characterized by a single mode of innovation, which correspond to the Type 4 of the Martin and Scott (2000) taxonomy: Applications of high-science-content technology, relying extensively on R&D as a source of knowledge.

4.7. Raising Capital

Focusing on the relationship between collaboration and raising capital reveals important characteristics. Regulations affect many of the new biotechnology products and processes and there can be significant time lags between invention and commercialization, creating in part a need for significant capital. Alliances can be a requirement of venture capitalists, signaling that firms seeking new capital are taking the necessary steps to gain core competencies. Alliances are one way of accessing that capability, knowledge and credibility (Stuart *et al.*, 1999; Greis *et al.*, 1995; Lunzer 1988). These various challenges mean different types of partners and partnerships. Collaborative arrangements were not done for money, but this must be qualified. An un-reportable number of firms cited collaborative arrangements as a source of capital. Firms were asked to directly attribute the amount of capital raised to a source and a confidential number cited collaborative arrangements as that source of capital.

However, viewing the results from another perspective uncovers other interesting information. The 95 firms that reported collaborative arrangements and successfully reported raising capital, raised over \$1.9 billion of the \$2.1 billion capital raised by 138 biotech firms that had successfully raised capital. This translates to collaborating firms raising on average \$20 million per firm compared to non-collaborating firms which raised almost \$4.7 million per firm. In fact 75% of the firms unsuccessful in their attempts to raise capital did not have collaborations. This corresponds with literature suggesting that collaborative arrangements give firms the credibility to raise capital.

Baldwin (1997) identifies two methods of research and development involving a strategic arrangement with another entity. Collaborative research "involves a partnership and, therefore, extends the boundary of the firm" and "is a substitute for contract research where third-party or market transactions do not work as well as internalization via the creation of a new entity" (p. 20). Contract research: "allows a firm to incorporate new ideas when it does not have external expertise. It works particularly well when the incorporation of new ideas and products does not involve tacit or firm specific knowledge" (p. 20). This type of alliance is market based. Biotechnology firms contracted out almost \$1 billion for a variety of purposes. By far the most common purpose was 187 firms contracting for \$858 million for research and development. An additional 85 firms contracted over $100 million for regulatory and clinical affairs and lesser amounts for marketing and management purposes. The human health sector contracted almost $400 million in research and development contracts and 98% of the regulatory and clinical affairs contracts.

5. CONCLUSION

Smaller firms tend to have slightly different collaboration patterns than large firms, but these observed differences for the most part are relatively minor and, when there are more significant differences, it may be related to another underlying factor. For instance, larger firms enter into more collaborations with universities than small firms. The data also suggest that the firms' performance of in-house R&D activities mostly explains collaborations between firms and public organizations such as universities, research institutes and government laboratories. Since more large firms perform R&D activities than small firms, this result may be more an internal commitment to R&D effect than a size effect.

The propensity to engage in collaborative agreements varies a lot by industry, ranging from 16% in the clothing industry to 61% in the pharmaceutical industry. The type of partner also varies. For instance, some industries like Petroleum and Coal products or Beverage and Tobacco are heavily engaged in relationships with their suppliers while others like Semiconductor Equipment or Pharmaceutical engage actively the universities.

The stage of development of the firm has a direct impact on the purpose for entering into collaborative arrangements. Over and above other collaboration purposes, small firms engaged in biotechnology are also using collaboration to access capital, markets, and distribution channels, and to protect intellectual property. Interestingly, the most important motive of large biotechnology firms engaged in collaborative arrangements is to gain access to R&D and specialized input.

Entering collaborative arrangements seem to have a direct effect on the success of the firm in raising capital. Or, perhaps is it more accurate to say that, if the presence of collaboration agreements is a prerequisite for obtaining venture capital, it also becomes a purpose for entering into collaborative agreements. If collaborative agreements are made solely for the purposes of obtaining venture capital, it may lead to a less robust alliance if the complementarities between partners are not present.

As the literature suggests, there are different modes of innovation. However, the taxonomies focus on industries and do not take into account the stage of development of the firm. As stage of development is a significant factor in determining the type of collaboration entered into by the firm, the existing taxonomies are not that useful for classifying types of collaboration.

There is also a measurement problem as there have been no attempts to incorporate a commonly accepted taxonomy of modes of innovation into the surveys of innovation which have been developed to support the comparability of the results. The development of classifications of modes of innovation and types of collaboration, and of any relationship between the two, remains a subject of future research.

Chapter Five
IN-HOUSE VERSUS EX-HOUSE:
The Sourcing of Knowledge for Innovation

Anthony Arundel and Catalina Bordoy[1]
MERIT, University of Maastricht

1. INTRODUCTION

Innovation researchers over the past decade have increasingly stressed the importance of external knowledge sources to the ability of firms to innovate. These include user-producer networks (Lundvall, 1992), collaborative research with other firms or universities (Hagedoorn *et al*, 2000) informal and formal links with universities (Mansfield, 1991; Pavitt, 1991), and contracted-out research (Howells, 1999). Several researchers have argued that innovative activities will increasingly depend on external knowledge sources (Gibbons, 1994; Georghiou, 1998), with Antonelli (1999) suggesting that "the systematic use of technological cooperation" will become "the dominant form" of producing new knowledge. Coombs *et al.* (2001) refer to "distributed innovation processes" in which many different firms and institutions are involved in an innovation. One of the main justifications for the increasing importance of cooperation is the belief that modern technology is growing in complexity and consequently beyond the abilities of a single firm (Kash and Rycroft, 1994). Under these conditions, firms must collaborate in order to develop competitive new products and processes.

The evidence in support of the importance of external knowledge sources to the innovative activities of firms is drawn from both case studies and from surveys, such as the first and second Community Innovation Surveys in Europe. Surveys consistently show that firms attach a high importance to information obtained from their customers and suppliers, from attending trade fairs and conferences, and from reading journals.[2] The survey results also point to two distinct types of external knowledge sources that firms find of value. The first group consists of external sources based on personal contacts, such as customers, suppliers, or cooperative research agreements. These are the types of external knowledge sources that require collaboration or cooperation between innovating organizations. The second group includes sources that can be accessed relatively inexpensively and which do not require personal contacts, such as reading journals, searching patent databases, or attending trade fairs.

Unfortunately, most innovation surveys suffer from a major drawback that reduces their value for investigating the role of external knowledge sources – there are very few questions on internal sources of knowledge, such as the firm's production engineering department, head office, related plants, sales and marketing staff, or R&D unit.[3] For example, the second CIS questionnaire includes ten questions on external information sources, but only two questions on internal sources within the firm or its group. The result is that most of the respondents' answers concern external sources, which, not surprisingly, means that most of the scholarly analyses and research papers highlight the importance of external knowledge. The nagging worry is that the apparent importance of external knowledge sources could be entirely erroneous and simply due to the design of the innovation surveys.

An alternative method for assessing the importance to firms of external knowledge sources is to use data on the location of R&D expenditures. These can include both expenditures to learn about research conducted outside the firm and contracted-out research. A 1997 survey of American R&D laboratories reports that firms spend less than 5% of their R&D learning about university research (Adams, 2001). Howells (1999) reports that contract R&D in the UK by business was equal to 10% of total business expenditures on R&D. These are comparatively low percentages of R&D spending, which contrasts with the subjective importance given to external knowledge sources in survey research.

One explanation is that many external sources of valuable knowledge are neither mediated by market relationships, such as contract R&D, nor expensive to acquire. For example, a valuable idea for an innovation can be obtained at almost no cost from a publication (Senker, 1995) or at a minor cost from attending a trade fair. Faulkner *et al.*(1995) report that the TRACE study in the UK found that half of the contribution of public research organizations to innovation was through firms reading the literature. Yet, new theories of innovation that stress technological cooperation or user-producer networks are not based on these quick and cheap methods of accessing knowledge. Instead, they stress long-term relationships that require expensive investments in time and in building up trust between partners. One possibility is that inexpensive public sources are used to obtain ideas for an innovation, but that more intensive personal contacts are used for the more difficult and expensive process of developing a good idea into an innovation.

Several factors will place an upper limit on the ability of firms to access external knowledge through cooperation or collaboration. Firms must develop innovative capabilities in-house, partly because of the sticky nature of knowledge (knowledge is not freely transmissible) and the need to coordinate different production activities, and partly in order to curtail rent-seeking behavior (Langlois and Foss, 1999). Otherwise, almost all firms could simply 'buy-in' all new technology, leaving only a few firms that develop innovations in order to earn license revenue. However, as firms develop internal innovative capabilities, they also increase their 'absorptive capacity' for external knowledge sources (Cohen and Levinthal, 1989). As an example, R&D investments give firms the ability to develop innovations in-house rather than relying on external sources, but simultaneously increase the efficiency with which they can apply external knowledge to their innovative activities. Several studies, using either theoretical or empirical approaches, find that the importance of external knowledge sources to innovation is mediated by the internal capabilities of the firm. (Cassiman *et al.*2001; Veugelers and Cassiman, 1999; Veugelers, 1997; Kamien and Zang, 2000)

These different perspectives on the value of external knowledge sources lead to a question that can be tested empirically – how important to firms are external knowledge sources compared to their own internal capabilities? We are particularly interested in external knowledge sources that require personal contacts. The answer to this question is of relevance to a wide range of policies that are designed to encourage firms to source knowledge externally. These include R&D subsidies, particularly in Europe, that require collaboration between firms or between firms and public research institutes, policies to encourage links between firms and public research institutes, and programs to improve access to external sources that do not require personal contacts, such as patent databases or scientific publications.

This study uses the results of the 2000 KNOW survey of small and mid-sized European firms to investigate some of these issues. The KNOW survey has several advantages over the CIS for this purpose. First, it includes a question on the percentage of innovations developed in-house, through collaboration, or via 'buying-in'. Second, the KNOW survey asks a series of detailed questions on the knowledge sources used by the firm to develop its most economically important innovation, following a similar technique used by Baldwin and Da Pont (1996). These include questions on the use of five different external sources for both the original idea behind the innovation and for help with its completion. Another question asks for an overall estimate of the relative importance of internal versus external knowledge sources to the development of this innovation.

2. METHODOLOGY

The KNOW survey was conducted in the spring of 2000 in seven EU countries: the UK, Denmark, the Netherlands, France, Germany, Italy, and Greece. Although the survey does not cover all EU countries due to funding limitations, the four largest EU economies were included plus two of the smaller, developed economies and one of the less developed economies.

The survey was limited to five sectors: food and beverages (NACE 15), chemicals excluding pharmaceutical (NACE 24 minus NACE 24.2), telecom equipment (NACE 32), telecom services (NACE 64.2), and computer services (NACE 72). These sectors were specifically chosen to represent low, medium and high technology manufacturing and to cover two innovative service sectors.

In each country, a random sample of firms from two size classes (10 – 249 employees and 250 – 999 employees) within each of the five sectors was drawn from a national business registry. A standard survey protocol based on a telephone CATI technique was used in all countries with the exception of the UK, where a postal survey was used. Table 1 gives the number of firms surveyed, the number of responses, the response rates, and the number of useable responses. The response rates by country vary from 9.6% in the UK to 76.5% in Denmark, with an average of 25.3% if the UK is included and 33.2% if the UK is excluded. The number of useable responses is less than the total number of responses because the analyses exclude non-innovative firms, firms that did not fit the sampling criteria for size and sector, and firms that did not answer the questions on their most economically important innovation. A maximum cut-off of 1,250 employees was used to allow for natural employment growth between the time that the data in the business registries were collected and the survey date.

Table 1. Survey Results by Country

	Firms Surveyed	Responses	Response rate	Useable responses
UK	1003	96	9.6%	44
Denmark	170	130	76.5%	78
Netherlands	331	151	45.6%	114
France	613	76	12.4%	65
Germany	470	101	21.5%	51
Italy	278	92	33.1%	75
Greece	260	110	42.3%	100
Total	3017	764	25.3%	527
Total excluding UK	2014	668	33.2%	

The response rates in Table 1 are based on all surveyed firms, although only 5.6% of the respondent firms did not innovate. If we assume that the target population only includes innovative firms, the estimated response rates increase substantially, since non-innovative firms can then be excluded from the population. For the Netherlands, limiting the target population to innovative firms only in each of the five sampled sectors increases the response rate from 45.6% to an estimated 72.7%.[4]

Some of the firms did not complete all survey questions. Where possible, the value of missing variables for interval variables were estimated using regression techniques based on the firm's sector, country, and size. Firms with estimated values are included in the descriptive results given below. On the contrary, they are excluded from the regression analyses.

The KNOW survey includes two question groups of relevance to the value of external versus internal knowledge sources. The first question asks the respondent to estimate the percentage of their product and process innovations, introduced within the previous three years, that were developed 1) "mainly in-house", 2) "in collaboration with external partners", and 3) by buying in, which includes "purchase, licensing, or contracting out development work". In-house development is restricted to the site of the firm, division, or subsidiary, while "external partners" includes other divisions of the same firm in a different location. This question provides a background measure of the firm's preference for innovating in-house or through external sources. The option for in-house development can also be interpreted as a measure of the firm's internal capabilities.

The second set of questions focus on the firm's most economically important innovation that it introduced in the previous three years. The key question is: "Overall, how important to the successful completion of this innovation were internal knowledge sources compared to external sources?" Three options were provided: 1) Internal most important, 2) External most important, and 3) Both of equal importance. As with the first question, external sources are defined to include sources within other divisions or units of the same firm, while internal sources must be located at the same physical site. This should increase the role of external sources compared to a definition based only on sources outside the firm.

Several questions provide additional information on the characteristics of the most economically important innovation and the knowledge sources that were used in its development. These include the type of innovation (product/service, process, combined product and process), whether or not the innovation had been patented, the receipt of government subsidies to develop the innovation, and the use of five different external sources as both a source of the original idea for the innovation and for its completion: competitors, suppliers, customers, universities or public research institutions, and consultants. Of note, the series of questions on the use of external knowledge sources does not include other divisions of the firm. The questionnaire asks if each of these five external sources contributed to the innovation on a yes or no basis. A second question then asks which of the five sources is the most important external source of ideas and the most important external source of knowledge for completing the innovation.

Other questions provide information on the firm. These include its ownership status (division, national subsidiary, or independent), the number of employees, sector of activity, and the firm's R&D status.

3. DESCRIPTIVE RESULTS

The majority of firms in the sample, 71%, have less than 249 employees, while 29% are mid-sized, with between 250 and 1,250 employees. The mid-size firms account for 78% of total employment among the sample. The majority of firms perform R&D, either on a continuous basis (71%) or on an occasional basis (25%).

3.1. Percent of all Innovations Developed In-House

Table 2 gives the average percentage, by sector, of all of each firm's product innovations that were developed in-house, via collaboration, or through buying-in. The majority of each firm's innovations, 70% on average, are developed in-house, with 16% developed through collaboration and 14% through buying-in. There are no significant differences by firm size in how firms innovate and only minor differences by sector, with one exception. Telecom service firms develop significantly fewer product innovations in-house than the average. Fewer process innovations (results not shown) are developed in-house (59%), with most of the difference, compared to product innovations, due to a higher rate of buying-in (23%) and a small increase in collaboration (18%).

Table 2. Percent of All Product Innovations (introduced in the previous three years) Developed:

	In-house	Collaboration	Buying-in
Food	68%	17%	15%
Chemicals	76%	12%	11%
Telecom equip	67%	17%	16%
Telecom services	50%	26%	25%
Computer services	73%	17%	10%
Total	70%	16%	14%

These results for all innovations provide a first fix on the relative importance of in-house versus external knowledge sources. Most innovations are developed predominantly through in-house activities. The results also show that almost half of externally acquired product innovations (and more than half of process innovations) are bought in. Collaboration, which by definition requires sourcing knowledge from outside the firm, accounts for a comparatively small share of innovations. However, collaboration is only one form of sourcing external knowledge. Some of the innovations developed mainly in-house could have depended on vital pieces of knowledge from external sources. Furthermore, these results are for all of each firm's innovations, with no adjustment for their economic importance to the firm. For both these reasons, most of our analyses of the role of external knowledge use the series of questions on the most economically important innovation that the firm introduced in the previous three years.

3.2. Most Economically Important Innovation

Several characteristics of the firms' most economically important innovation are given in Table 3, plus information on the use of external knowledge sources in its development. Product or service innovations predominate, with only 14% of small and 22% of mid-size firms stating that their most important innovation was a process innovation. On average, the product innovations accounted for 16% of the total sales of all small firms and 15% of the total sales of the mid-size firms, indicating hardly any difference by firm size.

The most important innovation has been patented for almost twice as many mid-size as small firms. These differences are not entirely due to sectoral effects, since the most important innovation is significantly less likely to be patented by small firms in the chemical and telecom equipment sectors compared to mid-size firms in these two sectors. However, the rates are not significantly different (although favoring mid-size firms) in food and beverages, telecom services, and computer services.

Table 3. Characteristics of the Most Economically Important Innovation

	Small (< 250 emps)	Mid-Size (250 – 1,250)
Type of Innovation		
Product/Service	55%	45%
Combined Product & Process	31%	33%
Process	14%	22%
Total	100%	100%
Sales share from product, service, product/process innovations[1]	16%	15%
Received a development subsidy	20%	18%
Patented (by firm or another organization)	21%	37%
External sources contributed to original idea[1]	79%	79%
External sources contributed to completion[1]	70%	77%

Almost equal percentages of small and mid-size firms (20% and 18%) received government subsidies to develop the innovation. Over 70% of firms noted that at least one of the five external information sources contributed to either the original idea behind the innovation or to its completion. The use of many of these external sources will depend on direct person-to-person contact and therefore raise the possibility of leaking information, in contrast to the use of sources such as reading the literature or accessing patent databases.

Table 4 provides a closer look at the use of each external knowledge source for the original idea or for completing the innovation. The table gives both the percent of all firms that cite any use of each source, plus the percentage that cite each as the 'most important' external source. The distribution of 'most important' sources sums to 100% of all firms. Some firms did not cite any of the five sources as 'most important', either because they used none of them or because a different external source was of greater value to them.

Customers are the most frequently cited source for the original idea, with 55% of firms citing any use of this source and 37% stating that it was their most important external source. Suppliers are the most cited source for completion (44%), followed by customers (40%). The same pattern holds for the most important source for completion, with suppliers in first place among the five external sources (27%) followed by customers (25%). These results support the literature on the importance of user-producer relationships, which stresses the role of users (customers) and producers (suppliers) to innovation.

Table 4. Percent of Firms Using Each Source for the Original Idea and to Help Complete Their Most Economically Important Innovation

	Original Idea		Completion	
	Any use	Most important	Any use	Most important
Competitors	37%	12%	10%	2%
Suppliers	35%	16%	44%	27%
Customers	55%	37%	40%	25%
Public science	12%	4%	14%	5%
Consultants	15%	6%	19%	9%
(None of the above)	-	25%	-	32%
Total		100%		100%

By sector (results not shown), the most relevant differences occur in the type of sources used for the completion of the most important innovation. The highest percentage corresponds to suppliers in all sectors except chemicals, where more firms cited their customers (46%) than their suppliers (38%). An above average percentage of respondents from the two service sectors cited consultants as their most important source (16% of telecom service and 12% of computer service firms). There are very few differences by firm size in the percent that find each type as their most important external source. An exception is that small firms are significantly more likely than mid-size firms to cite customers (40% against 29%). Also, mid-size firms attach more importance to suppliers and consultants than small firms.

3.3. Relative Importance of Internal Versus External Knowledge Sources

Table 5 gives the percentage of firms by sector that judged internal sources, external sources, or both equally to be of greatest importance for completing their most economically important innovation. The second part of Table 5 gives results by the receipt of government subsidies and for the type of innovation. The latter is divided into two groups: process innovations alone or innovations that contain a product component (this includes service innovations plus combined product/process innovations).

Our expectation is that the importance of external knowledge sources should be more valuable in high technology sectors that develop complex technology, such as telecom equipment, while internal sources could suffice for low technology sectors such as food and beverages, although external sources could be used in the latter for process innovations. The results for the telecom sector conflict with expectations, with 64% of telecom equipment firms finding internal knowledge sources to be of greatest value, which is significantly more than the average of 54%, while the results for the food and beverage sector are close to the average.

Table 5. Percent of Firms Finding Internal or External Sources to Make the Greatest Contribution to Completing Their Most Important Innovation

	N	Internal	External	Both Equally	
All firms	517	54%	15%	31%	100%
By Sector					
Food and beverages	123	52%	19%	29%	100%
Chemicals	125	54%	15%	31%	100%
Telecom equipment	92	64%	12%	24%	100%
Telecom services	45	51%	24%	25%	100%
Computer services	132	49%	11%	40%	100%
Received Subsidy					
Yes	95	43%	19%	38%	100%
No	327	56%	15%	29%	100%
By Type of Innovation					
Product innovation	427	54%	14%	32%	100%
Process innovation	82	51%	21%	28%	100%

The lowest reliance on external sources occurs in the computer services sector, where only 11% of firms find external sources to be more important than the alternatives. The cause is due to an above average percentage (40%) of these firms that find both internal and external sources to be of equal value. This could be due to the frequency of customization in this sector, in which software is developed to meet the customer's requirements. In support of this possibility, significantly more computer service firms than all other firms state that customers contributed to the original idea (63% versus 55%) for their most important innovation.

There is very little difference between product and process innovators in the preference for internal knowledge sources, but a higher percentage of process innovators prefer external sources (21% versus 14%), probably reflecting the role of equipment suppliers, while more product innovators find both sources of equal value (32% versus 28%).

Firms that received government subsidies to develop their most economically important innovation are considerably less likely to cite internal sources than firms which did not receive a subsidy (43% versus 56%), with an increase in both the 'external' and 'equal' options. This result could be due to the requirement, for many European subsidy programs, for firms to collaborate.

There are no statistically significant differences in the distribution of responses by whether or not the innovation was patented, by the R&D status of the firm (continuous versus occasional and never) the R&D personnel share, firm size, or ownership status (independent versus part of a group).

4. CLUSTER ANALYSIS OF KNOWLEDGE SOURCES

In this section we develop a cluster analysis in order to identify comparatively homogeneous groups of firms, based on ten questions: five for the type of sources that contributed to the original idea and five for the types of sources that contributed to the completion of each firm's most economically important innovation.[5] We chose a three-cluster solution as the most informative of use patterns for external knowledge.

Table 6 gives the percent of firms within each cluster that make any use each knowledge source. The most outstanding feature of firms in Cluster I, (termed 'contractors'), accounting for 16% of the firms, is a substantially above average use of public research (PR) and particularly of consultants. Many of these firms must contract consultants to solve innovation problems and they probably contract out problems to public research organizations, or conduct research on a cooperative basis. These firms are less likely than the average to patent their innovation (21% versus 26%) but more likely to perform R&D on a continuous basis (75% versus 70%). The second cluster (internal sourcers) includes 56% of the firms. These firms make below average use of external knowledge sources, particularly suppliers. Suppliers are cited by only 3% of these firms as a source of ideas (versus 35% for all firms) and by 16% for completion (versus 44% for all firms). Compared to the average for all firms, slightly fewer of these firms perform R&D on a continuous basis (67%).

Table 6. Results of the Custer Analysis

	Competitors	Suppliers	Customers	PR	Consultants
Cluster I (N=77): Contractors					
Idea	36%	47%	47%	30%	78%
Completion	14%	61%	35%	21%	87%
Cluster II (N=276): Internal Sourcers					
Idea	29%	3%	50%	8%	2%
Completion	5%	16%	27%	13%	6%
Cluster III (N=142): User-Producers					
Idea	50%	87%	66%	10%	5%
Completion	15%	87%	62%	12%	4%
All Firms					
Idea	37%	35%	55%	12%	15%
Completion	10%	44%	40%	14%	19%

The third cluster (user-producers) accounts for 29% of all firms. A very high percentage of these firms use suppliers and customers as a source of ideas and for help completing the innovation. They are considerably less likely to use consultants. These firms are significantly more likely to receive a subsidy for their innovation (28% versus 18% of contractors and 16% of internal sourcers). This suggests that subsidies encourage user-producer relationships more than they encourage firms to use consultants or public research organizations.

There are no significant differences by cluster in the type of technology (product or process) or by firm size. As expected, the majority of firms in the 'internal sourcers' cluster (66%) find internal sources to be of greater value in the development of this innovation than external sources or both equally. The lowest reliance on internal sources occurs among the 'user-producers', where only 36% find internal sources to be more important than the alternatives.

5. REGRESSION RESULTS

The regressions explore the factors that influence the probability that a firm finds internal sources, external sources, or both equally to be the most important knowledge source for the development of the innovation.[6]

5.1. Model Specification

There are three choices for the model specification, depending on the status of the 'equal' option. First, if the equal option lies in between the internal and external options in an ordered sequence, an ordered logit model is the best specification. Second, there could be no ordered relationship at all, with each of the three options being completely different choices. Under this condition, the best model is a multinomial logit. Third, the equal option could be indistinguishable from either the internal or external option. In this case, a simple binomial logit model would be best. Preliminary analysis showed that the binomial logit was the best specification. There is little difference between the two choices that include an external component: 'external' and 'equal'. Therefore, the dependent variable equals 1 when the firm finds internal knowledge sources to be more important than either of the two choices with an 'external' component.

The binary logit assumes that there is an unobservable variable y_i^* such that

$$y = 1 \quad if \quad y_i^* > 0$$
$$y = 0 \quad if \quad otherwise,$$

and y_i^* is defined by

$$y_i^* = \beta' X_i + \varepsilon_i,$$

where ε_i has a distribution function f derived from the logistic cumulative distribution function

$$F(x) = \frac{1}{1+\exp(-x)}.$$

Then, given the characteristics X_i of individual i, we have

$$\text{Prob } (y_i = 1) = 1 - F(-\beta' X_i) = \frac{\exp(\beta' X_i)}{1 + \exp(\beta' X_i)}$$

5.2. Independent Variables

The factors that influence the importance of knowledge sources for the firm's most economically important innovation will depend on general conditions within the firm and several firm characteristics, plus factors that apply specifically to this particular innovation. The former group include the innovative capabilities of the firm, firm boundary conditions such as its size and ownership status, its sector of activity, and its general strategies for sourcing external information. The latter group includes appropriation conditions for this particular innovation, the type of innovation, the types of external knowledge sources that the firm uses to develop it, and whether or not it received a government subsidy.

We use two variables as a proxy for the innovative capabilities of the firm. The first variable is the percentage of all product innovations that the firm developed in-house (see Table 2 above). We use product rather than process innovations because 84% of the most important innovations contain a product component. This provides a background measure of the rate of in-house innovation. A second measure of internal capabilities is the R&D intensity of the firm, measured as the share of all employees that are active in R&D.[7]

Small firms with less than 250 employees (SMALL = 1) could be less likely than mid-size firms (SMALL = 0) to seek external knowledge, since they have fewer staff that could develop the personal contacts required to access external knowledge.[8] Ownership status (independent or part of a larger firm) could also play an important role. The question on the relative importance of internal and external knowledge sources defines other divisions of the same firm as an external source. Therefore, independent firms (INDEPENDENT = 1) should be more likely than firms that are part of a larger firm (INDEPENDENT = 0) to rely on internal knowledge sources.

Internal expertise could suffice for the development of well-understood technologies, while complex technologies or technologies at the technological frontier could require firms to actively seek knowledge from external sources. Therefore, we expect the firm's sector of activity to influence the relative importance of internal versus external knowledge sources. Specifically, we expect firms in the high technology telecom equipment sector to place greater emphasis on external knowledge sources than firms in the low technology food sector. Each sector is entered into the regression as a dummy variable, with the food sector as the reference category. However, low technology firms could be more likely to outsource process innovations to specialized suppliers. To control for differences in knowledge sourcing by the type of innovation, the regression includes a dummy variable PRODUCT that equals 1 if the most economically important innovation is a product innovation. Process innovations are coded as zero.

In addition to using knowledge sources that require direct personal contacts, which raises the possibility of leaking strategic information, firms can use non-personal methods of sourcing external information. These include regularly reading the scientific and business literature to seek ideas for innovation (JOURNALS = 1) or similarly reverse engineering competitor's products (REVERSE = 1). These two variables apply to the firm's general innovation strategies and not specifically to its most economically important innovation. The first variable, JOURNALS, provides a measure of the general degree to which the firm searches for external knowledge sources of possible value to its innovative activities. We expect outwardly-looking firms to be less likely to depend on internal knowledge sources. Conversely, firms that frequently reverse engineer competitor's products should be less likely to need to rely on innovation via personal contacts with outside sources. For this reason, we expect REVERSE to increase the probability of depending on internal knowledge sources.

In addition to the variable PRODUCT, discussed above, the variables that specifically refer to the most important innovation cover whether or not it was patented, the use of specific knowledge sources, and whether or not the firm received a subsidy.

Strategic concerns over the release of information to competitors could influence the willingness of the firm to rely on external knowledge sources. Patents could play a role in reducing risk by conferring clear ownership rights and by reducing the probability of infringement. Firms could be more willing to rely on external knowledge sources when the most important innovation is patented, either by the firm itself (FIRMPAT) or by another firm (OTHPAT).

The dependent variable in the regression is derived from a question that refers to the overall importance of internal versus external knowledge sources to the successful completion of the innovation. For this reason, the use of each external knowledge source is limited to the most important source that 'contributed to the completion of this innovation'. The questions on external sources for the original idea are not included.[9] Each of the five external knowledge sources is entered into the regression as a dummy variable that equals 1 when it is the most important external knowledge source. The five variables are COMPETITORS, SUPPLIERS, CUSTOMERS, PUBLIC RESEARCH organizations, and CONSULTANTS. The reference category for external knowledge sources consists of firms that use none of the external knowledge sources or which report that none of them were the most important to them.

The final variable in the regression is a dummy variable (SUBSIDY) that equals 1 when the firm received a government subsidy to develop this innovation and zero otherwise. In Europe, many innovation subsidies, such as the EU Framework Programme, require firms to collaborate with other firms or with public research organizations. Therefore, we expect firms that have received a subsidy to be less likely to rely on internal knowledge sources.

5.3. Results

All regressions include dummy variables for country (Netherlands is the reference category), although the coefficients for the country dummies are not provided. The country dummies are included to adjust for national differences in the responses that are not captured by the other variables.

Table 7 (at the end of this chapter) gives the results of three different regressions. The first model does not include any of the variables referring to the use of external knowledge sources for the completion of the innovation. The second model contains two dummies for two of the cluster solutions, with user-producers as the reference category. The third model replaces the cluster solutions with the variables for each external source.

The results for the general variables (not linked to the most important innovation) are similar across all three models. The share of innovations developed in-house, a measure of both internal capabilities and the general 'background' preference for internal versus external sources, is highly significant and increases the probability that the firm relies on internal sources for this innovation. Conversely, neither the firm's R&D intensity nor its size has any effect. Unexpectedly, independent firms are less likely to rely on internal knowledge sources, indicating that they have a greater need for external knowledge than firms that can source external knowledge from other divisions of their firm. As expected, firms that regularly seek ideas for innovation from journals are less likely to rely on internal sources, but the use of reverse engineering has no influence on the outcome. The only difference by sector is for telecom equipment, which increases the probability of preferring internal sources. This result, although contrary to expectations, confirms the descriptive results in Table 5. However, the significance is lost in the complete model, where the sector effect is probably captured by the use of different external knowledge sources.

The remaining variables concern the firm's most important innovation. In all models, whether or not the innovation was patented has no effect on the preference for internal knowledge sources. The type of innovation (PRODUCT) also has no effect. Firms that received a subsidy to develop the innovation are significantly less likely to prefer internal knowledge sources.

As expected, the coefficient for the cluster 'internal sourcers' is positive and highly significant, in comparison to the reference category for 'user-producers'. In the complete model, all but one of the dummies for external knowledge sources reduce the probability of relying in internal sources. The exception is PUBLIC RESEARCH, which has no effect, although the coefficient is in the right direction. The lack of significance could be due to the small number of firms that find public research to be the most important external knowledge source (which reduces the probability of finding a statistically significant difference) or firms that rely on public research could still need to conduct a substantial amount of development work in-house. In contrast, firms that use suppliers or customers could rely on the latter for comparatively more assistance. None of the other four external sources (competitors, suppliers, customers, and consultants) are more important than the others for increasing the probability of preferring external knowledge sources, with overlapping 95% confidence intervals.

Table 8 (at the end of this chapter) provides binary logit results for four sectors with an acceptable model fit: food, chemicals, telecom equipment, and computer services. Separate results are provided by sector in case the assumption in the full model of linearity in the sector dummy variables is untenable. Due to a very small number of positive cases, a few variables could not be included in the sector-specific regressions. These include sourcing knowledge from competitors, public research for the food and telecom equipment sectors, and the use of journals by telecom equipment firms.

Patenting, product innovation status, and R&D intensity have no effect in any of the sector models. The receipt of subsidies significantly reduces the preference for internal sources in both the food and chemical sectors, but has no effect in telecom equipment and computer services, probably because very few of the latter received a subsidy. Independent food firms were less likely to prefer internal sources, but this variable is not significant in the other three sectors. Small telecom equipment firms are less likely to prefer internal sources. With regard to the methods used by the firms to obtain ideas for innovation, reading scientific journals has a negative effect on the preference for internal sources in the chemicals sector, but the effect is not significant for two other sectors, although the coefficient is in the right direction. The use of suppliers decreases the preference for internal knowledge sources in all the sectors except for the computer services sector. The negative effect of consultants on the preference for internal sources is limited to computer service firms.

6. CONCLUSIONS

The results of this survey show that small and mid-sized firms in five sectors developed 70% of their product innovations 'mostly in-house', with only 16% developed through collaboration with other firms and 14% obtained from buying the innovation from another firm. The low percentage of innovations that are developed through collaboration suggests that European firms in these five sectors are a long way off from the "systematic use of cooperation", as suggested by Antonelli (1999). Furthermore, the percentage of innovations that are bought-in is very similar to the percentage that are developed via collaboration. This shows that it is very important to be able to differentiate between the different methods that firms use to source innovations from external sources. Otherwise, we could mistakenly interpret buying-in for more complex collaborative activity between firms.

Although firms develop most of their innovations in-house, many of them could be of little importance to the firm. The results for the firm's most economically important innovation show a greater role for external knowledge sources, with 46% of firms reporting that external knowledge sources were either more important or equal to the importance of internal knowledge sources for its completion.

Competitors and customers are the most frequently cited sources for the original idea for the firm's most economically important innovation. The use of competitors probably occurs through non-personal contacts, as shown by only 12% of firms citing them as the 'most important' source of the original idea, compared to 37% that cite customers. Suppliers and customers are the most frequently cited sources for completing the innovation, which supports the importance given in the literature to 'user-producer' relationships. However, the cluster analyses show that only 29% of the firms rely on user-producer contacts, with 56% of the firms making little use of these sources, with the exception of customers as a source of the original idea. The remaining firms, 16%, place greater emphasis on contract organizations such as consultants or public research institutes.

We expected firms to be more likely to depend on external sources for developing process innovations, since they were more likely to be 'bought-in' than product innovations. However, there is no significant difference, either in the descriptive or regression results, in the percentage of process and product innovators that find external knowledge sources to be more important than internal sources.

The regression results show that one of the most important factors in the preference for internal versus external knowledge sources is the firm's background rate of in-house innovation. Given this rate, the regressions evaluate the effect of other factors on the relative importance of internal and external knowledge sources. Surprisingly, several factors that have been cited in the literature as influencing the use of external knowledge sources have no effect. These include the firm's sector of activity, the firm's R&D intensity (measured by employee share), and whether or not the innovation was patented.

We expected firms in the telecom equipment sector to be less likely to develop innovations in-house due to the complex nature of telecommunications technology. The firm's R&D intensity could also be positively correlated with the importance of external knowledge sourcing, for similar reasons. However, the results suggest that telecom firms are more rather than less likely than other firms to rely on internal technology. A possible explanation is that firms active in low technology sectors rely on their suppliers for process innovations, thereby confounding the results, but this is an unlikely explanation, since the regression models include a variable to control for product or process innovations. An alternative explanation that can't be addressed in this study is that technological complexity in the telecom sector only affects the knowledge sourcing strategies of large firms with over 1,000 employees.

R&D intensity has no effect on the relative importance of internal and external knowledge sources, which further suggests that technological complexity has little effect on the relative importance of internal and external knowledge sources. We must point out, however, that the question refers to relative values. Increasingly complex innovations could require increasingly high levels of internal expertise.

Patenting could increase the probability of using external knowledge sources by solving ownership disputes for intellectual property. But, whether or not the innovation was patented had no effect on the importance of internal versus external knowledge sources. This is possibly because appropriation issues and concerns over information leakage play only a minor role in the decision to obtain information from external sources.

Two factors significantly reduced the probability that the firm would rely on internal versus external knowledge sources. First, firms that receive innovation subsidies are considerably less likely to prefer internal sources. This result could simply reflect the requirement to collaborate in order to receive a European subsidy. Of greater concern, this suggests caution in interpreting survey estimates of the frequency with which firms collaborate, since these estimates will be influenced by policy. This will make it more difficult to determine if firms would willingly enter into collaboration in the absence of subsidies. Second, firms that regularly search the scientific and business literature for new ideas for innovation are less likely to rely on internal sources for their most important innovation. This effect is particularly strong among firms in the chemical sector. These firms could use the literature to identify potential partners for developing an idea into an innovation.

Table 7. Binary Logit Results for Internal Knowledge Sources as the Most Valuable (Reference to External/Equal)

	Reduced Model	Model with Clusters	Complete Model
Constant	-.104(.52)	-.39 (.569)	.91 (.596)
General Variables (not linked to the most important innovation)			
Innovative Capabilities			
% Products Developed In-House	.017***(.003)	.016***(.004)	.018***(.004)
R&D Employee Share	-.003 (.004)	-.001 (.005)	-.006 (.005)

Table 7. Binary Logit Results for Internal Knowledge Sources as the Most Valuable (Reference to External/Equal) (Continued)

	Reduced Model	Model with Clusters	Complete Model
Firm Boundaries			
SMALL	.24 (.27)	.03 (.286)	.28 (.291)
INDEPENDENT	-.57**(.263)	-.59**(.278)	-.55**(.278)
Outward-Looking Approach to External Knowledge Sources			
JOURNALS	-.61**(.289)	-.66**(.305)	-.53*(.311)
REVERSE	-.21 (.245)	-.18 (.26)	-.05 (.268)
Sector			
Chemicals	-.15 (.318)	-.29 (.334)	-.22 (.338)
Telecom equipment	.71*(.355)	.72*(.377)	.62 (.386)
Telecom services	-.32 (.461)	-.46 (.493)	-.34 (.506)
Computer services	-.25 (.316)	-.28 (.331)	-.35 (.34)
Variables Limited to the Most Important Innovation			
Patented			
FIRMPAT	.29(.307)	.16 (.329)	.25 (.456)
OTHPAT	-.49(.371)	-.59 (.394)	-.54 (.414)
Technology Characteristics			
PRODUCT	.01 (.298)	.08 (.313)	.23 (.328)
Cluster Solution Variables			
CONTRACTORS		.54 (.363)	
INTERNAL SOURCERS		1.2***(.277)	
Most Important External Knowledge Source for Completing the Innovation			
COMPETITORS			-2.9**(1.162)
SUPPLIERS			-1.86***(.327)
CUSTOMERS			-1.8***(.344)
PUBLIC RESEARCH			-.79 (.596)
CONSULTANTS			-1.54***(.451)
SUBSIDY	-.69**(.298)	-.64**(.32)	-.71**(.335)
Number of cases	434	417	430
Pseudo R-square	.224	.289	.338
Model chi-square	79.659***	101.574***	125.5***
% correctly classified	68	71.5	72.3

* p<0.1, ** p<0.05, *** p<0.001 Standard Errors in brackets

Table 8. Binary Logit Results for Four Sectors for 'Internal' Knowledge Sources

	Food	Chemicals	Telecom equipment	Computer services
Constant	3.48** (1.662)	2.47*(1.487)	-.607*(1.859)	-.27 (1.309)
General variables (not linked to the most important innovation)				
% products developed in-house	.017*(.009)	.03**(.01)	.02*(.011)	.015 (.01)
R&D employee share	-.015 (.012)	-.01 (.01)	.002 (.014)	.002 (.009)
Firm boundaries				
SMALL	.56 (.62)	.78 (.698)	-2.06**(1.042)	.35 (.738)
INDEPENDENT	-1.78**(.744)	-1.14 (.714)	.84 (.988)	-.11 (.648)
Outward-looking approach to external knowledge sources				
JOURNALS	-.69 (.773)	-2.17**(.855)		-.91 (.72)
REVERSE	-.64 (.75)	-1.13 (.799)	.45 (.747)	.3 (.542)
Variables limited to the most important innovation				
Patented				
Innovation patented	1.26*(.783)	-.42 (.631)	-.92 (.75)	-.44 (.666)
Technology characteristics				
PRODUCT	-1.02 (.846)	.73 (.676)	1.06 (1.01)	.53 (1.008)
Most important external knowledge source for completing the innovation				
SUPPLIERS	-2.39**(.811)	-1.54**(.772)	-1.78**(.853)	-.91 (.682)
CUSTOMERS	-1.76**(.781)	-1.12 (.779)	-1.83**(.897)	-2.15**(.788)
PUBLIC RESEARCH		.07 (1.068)		.08 (1.374)
CONSULTANTS	-1.37 (1.138)	-1.02 (1.208)	-.65 (3.138)	-2.61**(1.054)
SUBSIDY	-1.85** (.772)	-1.87**(.905)	.47 (1.153)	-.21 (.828)
Number of cases	100	108	80	110
Pseudo R^2	.46	.468	.497	.418
Model chi-square	42.204***	46.724***	35.819***	41.326**
% correctly classified	78.0	78.7	78.8	72.7

Notes: All analyses include country dummies, with the Netherlands as the reference category. * = .05 < p < .10, ** = .01 < p < .05, *** = p < .01. Standard Errors in brackets

PART III

PRACTICE

Chapter Six
INNOVATION THROUGH LINKAGES AND NETWORKS AT THE NATIONAL RESEARCH COUNCIL:
Measuring the Output

Jacques Lyrette,
Vice-President – Technology and Industry Support
National Research Council of Canada

1. INTRODUCTION

The Conference Board of Canada defines innovation as: "a process through which economic value is extracted from knowledge through the generation, development and implementation of ideas to produce new and improved products processes and services." The innovation process requires investments in a number of activities that must be linked in dynamic and creative ways in order to produce desired outcomes: wealth and prosperity. This new setting has affected the way they go about their business, reshaping our social, cultural and organizational entrepreneurship behavior. In this context, networks have become the instrument of choice in the innovative economy. They provide a platform, to maintain connections, to move within and between organizations, to build relationships, to share knowledge, etc. Networks thus make it possible to create intellectual capital in a collaborative and virtual way.

There have been many debates on the management of knowledge and the measurement of intangibles such as intellectual capital in the context of innovation. There is a growing interest in how organizations and firms acquire, measure and manage their intangible assets, such as intellectual assets. While accounting bodies are still wrestling with how such assets should be reported, there are a number of pioneering companies who have already developed appropriate measures and are using them in their goal setting and management practice. But it is nearly impossible to value and access wisdom embedded in people, the most important ingredient of the innovation process. Gains in human performance and competitiveness remain difficult to measure, which makes it difficult to value leveraging of intellectual capital and know-how that can be found in linkages and networks.

The return on investment (ROI) on interactions between people (linkage and networking activities) is difficult to measure because the nature of the process and the attribution of the result raises a number of questions. *How to share the benefits resulting from such a process because the contributors are distributed and it is near to impossible to measure the contribution of individuals*? How do you determine that an *innovation occurred* quicker because of an interaction that was stimulated by an individual in a network? And if you can determine that, how can you measure the tangible gain from a highly intangible activity? These things can lead to executive-level discomfort with the costs of investing in networks and linkage activities. Scientists, *of course, have been practicing such activities* - networking and linkage - in the normal process of creating new knowledge, which may benefit the community as a whole, as well as individual researchers.

The key factor for higher return on an investment in networking is the effective utilization of information and knowledge as it relates to the stated objective. Is there a relationship between investments in networks and/or linkage activities and the performance of a program, or the creation of new knowledge or a new product? It can be argued that the real value of networks is linked to a better understanding as to the process of how knowledge flows, and how collaboration and sharing take place. Building on each others knowledge, for whatever reason, are other key activities. This is what people networks are all about. Instead of creating a competitive edge through information, they create a competitive edge through knowledge sharing and knowledge generation.

This chapter considers three examples that show the value of networking and linkages in fundamental research, industrial materials research, and role of the Industrial Research Assistance Program (IRAP), which has made a difference for university spin-offs.

2. THE INNOVATION PROCESS IS A VERY INTERACTIVE ENVIRONMENT[1]

Invention implies something new, whereas *innovation* refers to something both new and practically applied: in other words, useful in some fashion to someone. The term *invention* is often associated with radical discoveries, the transistor and the laser being excellent examples. But there are all kinds of inventions, patentable and otherwise. Some are extremely simple, while others very ingenious, and some are very complicated. An innovation does not have to come from a radical invention to be successful. It may be an ingenious way of putting things together, or it may incorporate something new that gives customers more value for their money. Or, it may be a complex combination of technological components that give rise to a superior new product, such as a very powerful hand held computer with all kinds of utility features.

The term *innovation* is, in fact, difficult to define accurately except for its connotation of newness, utility and competitive marketability. Even "newness" is a relative term: to what extent is the new product actually "new" for the firm, for the region, for the country, or for the world? What matters is that it is a departure from the current: it sells, it makes a profit, it out-does competitors, it establishes itself in the marketplace, and it is recognized as useful and significant. Successful innovations are critical factors in the growth of advanced economies; this growth, in turn, is often linked to social progress.

In the past, most public policy has relied largely on the traditional linear model of the innovation process, which leaves much to be desired. This model starts off with basic research, which leads to inventions and then to applied research, followed by experimental development and finally commercialization and industrial exploitation of new products and services.

As de la Mothe (1994) has pointed out, there are numerous flaws in the linear model. Not only is it highly stylized, it does not reflect reality. It does not accurately reflect the way industrial research laboratories or businesses operate. Innovation encompasses far more than just R&D, and the model does not take into account the social, organizational, market and financial factors upon which most new products and economic growth are based. Industrial researchers and product/service innovators can hardly identify themselves with this framework. In real life, industrial innovation does not result from a series of hermetically-sealed activities - basic research, applied research, development, etc. - organized in a linear sequence. The innovation process is not a vague grouping of territories of scientists, engineers, accountants, financiers and marketers, each with discrete, rather inflexible and hardly transferable skills. Instead, it is an integrated human resources framework, which is often referred to as a people network.

In reality, this process of successful innovation is far more complex. It is a series of iterative steps, as well as interplay of different elements of the innovative system: public sector, private sector, universities, manufacturing, markets, financial community, users and consumers. This complex process and the interplay amongst the different elements of the process represent the essence of linkages and networking.

Simply stated, the innovation process links all the necessary scientific, technical, commercial and financial elements needed for the successful development, production and marketing of new and improved products, processes and services. It is far more than just finding a practical use for an invention or translating R&D results into some marketable object. It is a very complex system that embraces both the public and private research both fundamental and applied, the financial community and governments. Figure 1 illustrates the links between the various parts of the system.

Figure 1. The Conceptual Framework for Canada's Innovation System

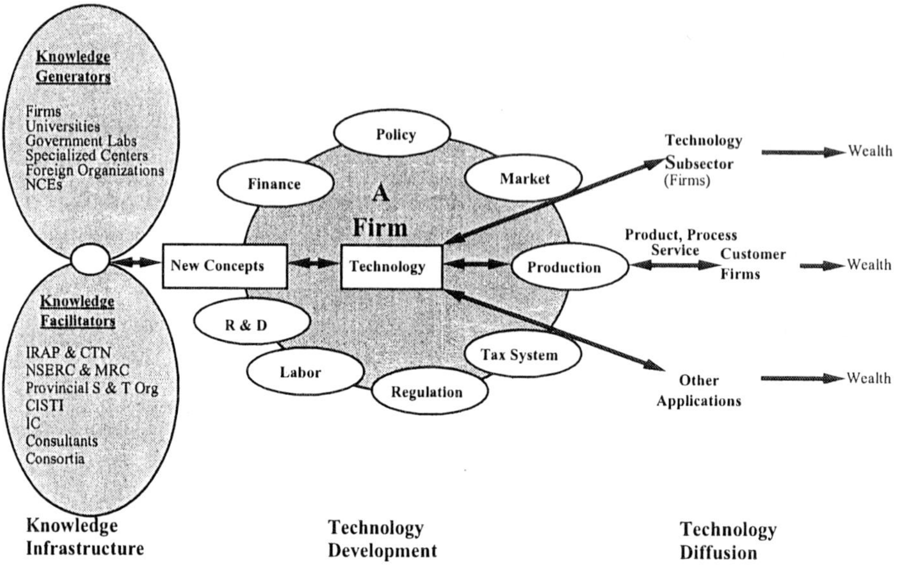

Source: NRC, Ottawa, April 1996

2.1. The Linear Model of Innovation

Simply stated, the linear model is:

Basic research —> applied research —> development —> innovation —> design —> pre-production —> commercialization

This is the model that prevailed in the 1960s, and was used in national science policy discussions when attempts were made to justify larger investments in fundamental research. Unfortunately, it is overly simplistic and does not apply to a wide range of technologies, especially complex technologies. However, because it still lingers in the minds of many decision-makers, a few clarifications are in order.

Essentially, the linear model is a science-push paradigm. It entails broad-based, undirected basic research that eventually brings about major discoveries, which eventually end up as new products and new industries. According to this ideology, basic science discovers new principles and new mechanisms. When a practical application is thought of — further applied research explores its possibilities. This leads to experimental development, engineering and industrial design, prototyping, beta site testing, industrial pre-production and ultimately market launch. In this model, knowledge is discovered in universities; it is passed on to industry through publications, patents, graduates and other forms of technology transfer, and on to final customers as new products or services.

2.2. The Feedback Loop Model

A more satisfactory model of innovation, illustrated in Figure 2, consists of a series of feedback loops connected in a chain-linked fashion. It portrays a complex series of inter-related activities with numerous feedback mechanisms involved.

In this framework, knowledge and research are not dissected as they are in the linear model. Competitive and pre-competitive research fall in line naturally, and so does experimental development with regard to research per se.

Importantly, the innovation process is closely linked with the body of knowledge, skills and experience existing within the firm (core competencies) on the one hand, and the process of acquiring new knowledge — and answers to problems (scientific research) — on the other. Also, experimental development is closely linked, as it should be, to both research and industrial/engineering design and production.

The key element of this paradigm is the recognition of an invention and analytical phase, which is based, at its outset, on a specific or generic market need. Contrary to the science push or technology push model, where too often "new products are seeking practical applications," the new object or artifact finds, here, a ready application and demand.

Early recognition of an important problem or a special contract opportunity is also fertile ground for productive inventiveness and successful innovation. This is especially true when the solution contemplated is fully compatible with the core competencies of the firm.

Figure 2. The Feedback Loop Model of Industrial Innovation

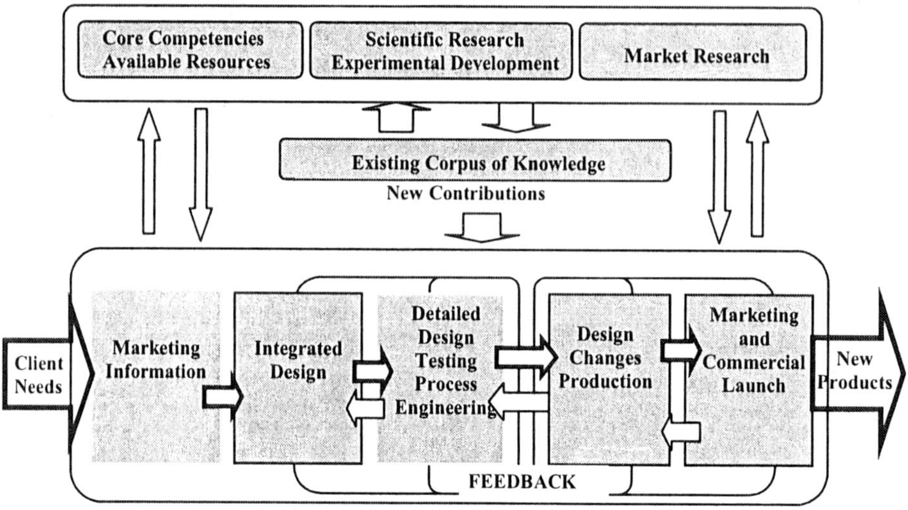

Source: OECD, National Systems of Innovation, 1994

Most successful innovations are based not on radical inventions, but rather on the interplay between various technological, commercial and socio-economic factors. Clearly, invention and innovation alone are not sufficient for a successful innovation system. The diffusion of knowledge and know-how is also critical to a successful innovation system. Thus, it is people who make the difference and specifically, people who have both knowledge and experience: who know how to do things.

The innovation process is an intensely human endeavor that must interact in a very dynamic way and relies on strong relationships amongst all stakeholders.

3. DEFINITIONS[2]

As indicated above, *innovation* is a serial process, a highly interactive process that relies heavily on the relationships between people with clear objectives, trust and knowledge.

Webster's New Collegiate Dictionary defines knowledge as "the fact or condition of knowing something with familiarity gained through experience or association." This definition of knowledge is universally used and accepted, but it begs the question as to the meaning of the term "knowing". "The word "know" is defined as "to perceive directly: have direct cognition of." and cognition is defined as "the act or process of knowing including both awareness and judgment." So knowledge implies cognition, and cognition implies awareness. And computers, quite obviously, do not have "awareness"—at least not yet.

Information, on the other hand, adds context through relationships between data, and possibly other information. Data with meta data and context makes information. The relationships may represent information, yet the relations do not actually constitute information until they are understood. Also, the relationships that represent data have a tendency to be limited in context, mostly about the past or present, with little if any implication for the future.

Knowledge adds understanding and retention to information. It is the next natural progression after information. To have "knowledge" requires information in conjunction with patterns between data, information, and other knowledge, couples it with understanding and cognition. Types of knowledge include "knowing about", "knowing why" and "knowing how".

Wisdom can be thought of as knowledge applied but not just 'knowing how" because it is obtained through experience and life long learning and is gained through experience and association. Wisdom is embedded in people.

Networks bring together all of the above mixed together in a very loose association, both formal and informal that transcend the boundaries of organizations and countries which come together to create a virtual community of people, practices and objectives. But how much of each is needed depends highly on the nature of the innovation, the issues that must be addressed and of course the quality of the people contributing to the network.

4. THE VALUE CHAIN

Knowledge, information and wisdom are distinct entities. All are key to the innovation process.

Information can be codified and handled by computers but they do not have awareness, which is a combination of knowledge and wisdom. The highly distributed nature of knowledge and wisdom and the speed of change in the knowledge based innovation process, give rise to the creation of networks and of linkages. The gathering of the collective knowledge from all of these sources can only be achieved through networking and linkages because it is highly distributed and too costly to generate such knowledge in highly competitive global markets. This is especially true for SMEs, which represent close to 99% of Canada's industrial infrastructure. Of course large firms can afford to have access to knowledge and wisdom to establish their competitive edge, but in the face of globalization and rising costs, they also turn more and more to networks to acquire knowledge and competitive intelligence.

Figure 3. The Creation of Knowledge through the Interaction of Information, Knowledge and Wisdom. – The Value Chain

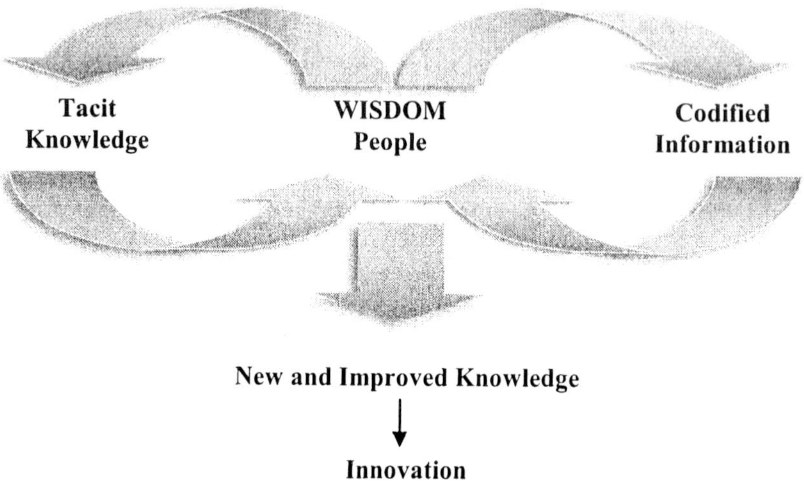

Information is not very rich in human interpretation for potential action, knowledge resides in the user's subjective context of action based on that information and wisdom is acquired through experience. Knowledge and wisdom reside in people and not in the collection of information. Knowledge caters to the critical issues of adaptation, survival, and competition in a discontinuous environment. Essentially, it embodies organizational processes that seek synergistic combination of information technologies and the creative and innovative capacity of human beings. In other words, knowledge is embedded in people and knowledge creation occurs in the process of social interaction. At the present time, only human beings can take the central role in knowledge creation and diffusion that result in the creation of intellectual capital in firms and organizations.

Creating the environment that encourages the organizational processes so that people can associate and interact with the view of achieving common objectives is what networks are all about. The result is linkages amongst people that have the knowledge to solve issues by providing incremental knowledge so that new knowledge can be created. In other words, the result is more than the sum of the parts.

But the issue here is that the intellectual capital is distributed amongst a number of people and organizations within the network and can only be sustained by ensuring that new knowledge is generated, shared and used. The high value of knowledge stems from its exploitation, which provides the ROI to sustain knowledge generation. It is therefore difficult to put a value on role and scope of networking and its result. Should this value chain be broken, the network would disappear and new ones will be created as part of the normal evolutionary process of knowledge based activities.

5. CANADA A NATION OF NETWORKS

Because of its geography, its socio-economic environment, its political system and the distributed nature of its innovation systems, Canada, had to rely on networks to link its institutions and increase its competitiveness through innovation. During the past decade, Canada has made remarkable strides in developing a number of national networks that allow people from across the country to work together on important issues and obtain and distribute information, create new knowledge and promote innovation.

The Networks of Centres of Excellence (NCEs) bring researchers from across Canada to create new knowledge cooperatively in fields identified as vital to Canada's future. Their output is the development of highly qualified personnel (HQP). They have been credited[3] for the development of over 3500 HQP: 992 university researchers, 184 non-university researchers, 1208 directly supported HQP and 1349 supported by other sources.

The Canada's Advanced Network for Research, Industry and Education (CANARIE), now CANARIE Inc., was established in 1993, with federal and private sector involvement, to facilitate and speed up the development of critical aspects of Canada's communications infrastructure. CANARIE funded the original Internet backbone in Canada; its most recent project has been the CA*net 3 network - the world's first national optical R&D network. This will be extended by a further commitment of $110 million over the next 2 years to build CA*net 4. CANARIE's present focus is to accelerate the development of Canada's Internet and to facilitate the creation of innovative applications that exploit the power of the most advanced Network in the world. During the period from 1993 to 2001, CANARIE is credited with having connected over 80 universities, over 50 community colleges, over 2000 schools, 10 research centers, 40 countries, funded over 200 projects, participation of over 500 companies and having created thousands of jobs.[4]

Precarn is a national consortium of corporations, research institutes, universities and government partners working with the Intelligent Systems industry. Precarn's focus is to make Canadian firms more globally competitive through increased development and use of intelligent systems technologies and expertise. Since its founding in the 1980s, Precarn has engaged over 1,200 individuals and 100 companies in a suite of major research projects at the leading edge of intelligent systems research.[5]

The government of Canada has created a number of networks, such as the Connecting Canadians program and others, which have kept Canada among the leaders in connecting its citizens to the Internet. The program has given Internet access to schools, libraries (SchoolNet) and communities across Canada (Community Access Program) so that Canadians, no matter where they are geographically situated, could benefit from the information and tools available through communications technologies. As of May 2000 there were close to half a million computers in Canadian schools with access to Internet and links more that 5,000 teacher approved learning resources, making it easier for Canadian teachers and students to access an immense source of online resources.[6] Canada Online is helping connect Canadians to each other and the world, by establishing 10,000 public Internet access sites in rural, remote and urban communities including schools and libraries. Even more recently, the federal government proposed spending $600 M over the next three years for Government On-Line to provide Canadians direct access to government service.

Along with other federal partners, NRC established the Canadian Technology Network (CTN), a network of over 1 000 individuals throughout Canada to assist SMEs with technology advice, expertise and intelligence. CTN, in turn, complements NRC's Industrial Research Assistance Program, another network that comprises some 260 Industrial Technology Advisors based in every province of Canada. Through its Canada Institute for Scientific and Technical Information (CISTI), NRC provides Canadians with access to a world-class collection of science, technical and medical information through an electronic network. NRC created the Canadian Bioinformatics Resource as an integrated and pan-Canadian facility to provide Canadian researchers with access to the global genomics database and bio-informatics tools. NRC's recent efforts in promoting the development of community-based technology clusters are another example of networks at the local and regional level creating value for Canadians. NRC is still trying to find output measures for these network activities in the sense of measuring their ultimate impact.

These networks are significant elements of the Canadian innovation systems. They help to overcome Canada's geographical and demographic realities, enable to link Canadians and to share information and knowledge and combine their ideas and capabilities in response to national challenges. They have helped Canada move from 10th place to 7th in the recent international competitiveness ranking by the Washington-based *Foreign Affairs* magazine.[7]

These networks help advance Canada's socio-economic agenda. They bridge communities of interest to come together to tackle issues of concern. They also provide the means and opportunity to bring disparate parties together on wide-ranging perspectives on challenges and develop innovative approaches and solutions. But the real question remains: what tangible evidence do we have that the ROI for Canada has justified or justifies the investments?

The Canadian innovation system is fragmented and can best be described as a mosaic of regional innovation systems and not a seamless national system. Interestingly, in order for Canada to maximize its innovation performance and be more competitive in global markets, Canada must increase its ability to nurture existing networks and create new ones. Enabling more Canadians to work with one another certainly encourages social cohesion but more importantly, it draws maximum value from the distributed assets. NRC has built regional networks and NRC's IRAP / CTN helps connect them nationally. Nevertheless, measuring their outputs in terms of economic benefits is still a challenge so that they are seen as a key instrument of our innovation system.

6. MEASURING OUTPUT

Although Canada has been very active in creating and establishing networks, measuring the ROI is still difficult and very empirical, to say the least, and can only be done at a fairly macro level. Most of the quantifiable outputs have been based on the elements of the networks that can be quantified (such as the number of firms and schools connected, or the speed of access, etc.) rather than the less tangible value that these networks bring to support the innovation process. At the most fundamental level, these values amount to concepts such as attribution: who is responsible for making the difference and what is the individual contribution of the network members?

What are the appropriate measures for these intangibles such as networking and creating linkages or relationships? First, if one looks at the different elements of the value chain it is necessary to understand the different components that constitute intellectual capital. An increasingly popular classification divides intellectual assets into four categories:[8]

1) Human Capital (knowledge and wisdom) – "what goes home at night", -: knowledge, competencies, experience, and know-how within the minds of individuals, and contribute to wealth generation.
2) Structural Capital - "that which is left after people go home for the night"; the inanimate processes, information systems, and databases etc. that exist within organizations
3) Network Capital – relationships
4) Intellectual property - inventions, patents copyrights, trademarks and licenses.

Creating networks is often confused with the creation of physical infrastructures or intellectual property: information technologies to build networks (structural capital) that create the knowledge so critical to business activities and those investments can be traced to such items as dollar value, number of patents, and licenses etc. These are quantifiable assets. But the real value of networks resides in the people interaction and the relationship building, and not in the physical platform that support the process, the human capital, which is difficult to measure. It is often suggested that increasing investments in networking should somehow result in improved competitiveness and increased innovation. Investments in networks have to be seen as improving the performance of organizations and firms or the delivery of improved information through the generation of new and improved knowledge. Unfortunately, tangible assets are often used as the basis for making decisions on the ROI for investing in networks because the generation of intellectual capital through networking and the building of linkages are difficult to measure. It also takes a long time for tangible benefits to be seen.

The creation of human capital through the building of relationships embodies organizational processes that seek synergistic combination of information technologies (structural capital) and the creative and innovative capacity of human beings (human capital). In other words knowledge is embedded in people and knowledge creation occurs in the process of interaction (relationships). Networks are about capitalizing on the intangible human assets, captive minds, experiences, and relationships. Establishing the value proposition should focus on measuring the output of such activity.

7. THE NATIONAL RESEARCH COUNCIL OF CANADA

The NRC is a knowledge and innovation organization that spans the innovation spectrum, from discovery at the frontiers of knowledge, to applied engineering, to technology commercialization. It addresses critical national issues and helps position Canada to respond to future opportunities and emerging challenges. NRC is unique in its seamless integration of strategically focused research, national standards activities, industrial infrastructures, scientific and technical information programs and industrial research, advice and innovation services.

As an organization with global, national, regional and local presence, NRC relies on a broad network to other innovators and mobilizes this network in response to national goals and priorities. NRC is present in over 90 communities across Canada, with a network of 19 institutes, six technology and innovation centers, 17 NRC science and technology Information Centres, 260 industrial technology advisors in its Industrial Research Assistance Program (IRAP) and more than 1,000 Canadian Technology Network members. Its presence stretches even further through thousands of national and international partnerships, collaborations, committees and networks.

NRC has been central to the Canadian innovation system at the international, national, regional and local levels. It can be seen as an integrator of the system, from discovery to innovation, and relies heavily on networking to create value for Canada. NRC creates and exploits knowledge to the benefit of Canada in a number of ways:

Direct technology transfer industry- The conduct of research (fundamental and applied) leads to research findings that may eventually find applications and there may ultimately be economic benefits from application of these findings. The results can also improve the quality of life of Canadians but some research findings usually take years or decades before any benefits result.

Training of people- Because of contact with leading-edge technologies employed in research it is highly likely that there will be economic benefits associated with the training of graduate students and postdocs who eventually go on to work in industry. These individuals will have been exposed to leading-edge thinking, instrumentation, and software related to a number of technologies.

Technical assistance to industry- As a result of the process of carrying out research and development, NRC develops a wide range of competencies and infrastructures. These are used to provide advice and assistance to industry on their own technology development and application, entirely separately from any direct technology transfer. In addition the R&D activities lead to privileged access to knowledge across the world and a source of competitive intelligence.

International business opportunities for industry- As a result of conducting world-class research, and being involved in international consortia NRC develop credibility with the world community. This can be used to identify and promote international business opportunities for niche products, services, etc. and even develop expertise in the construction of very sophisticated facilities.

Direct support to industry- NRC provides direct assistance to industry through the IRAP and the Canada Institute for Scientific and Technical Information (CISTI). IRAP relies heavily on the broad networks of advisors across the country and the network of NRC institutes to provide advice and financial support to Canadian SMEs. IRAP also provides assistance to SMEs to access technology abroad or to develop linkages with similar organizations across the world. On the other hand CISTI provides S&T information to the whole of the Canadian innovation system (public and private sectors).

In carrying out these activities, NRC relies heavily on networking activities to create new knowledge and increase its intellectual capital. This capital is created through a number of networking activities: university collaborations, industrial collaboration, fee for service, international collaboration, industry support, conferences, etc. NRC has been focusing on the issue of measuring the value of its activities and has developed performance measures with the objective of establishing its contribution to Canada's wealth. Previously, most of the measurements were based on the traditional measures, such as publications, the number of collaborations, the financial value of in-kind contributions, etc. These are activities that we can count. Recently, NRC has been looking at other measures such as the spin-off firms, the industrial benefits of its activities, the pay back of its international activities, etc. by trying to capture the value of the results of its network-based activities.

Three examples have been selected to illustrate the sorts of value that Canada obtains from NRC's network-based activities. The first example is in the fundamental science, the second is in industrial research, and the third is in industrial support to SMEs.

8. FUNDAMENTAL RESEARCH

Research, even of a very fundamental nature, has been shown by many recent studies to have substantial impacts. But researchers have over the years been early adopters of the concept of networks and linkages and have exploited these means to generate new knowledge. This is especially true with the advent of Internet, the globalization of science (e-science or e-research) and of innovation. In addition, most of the fundamental science either involves the use of large scientific infrastructures that are built though multinational consortia or through broad networks of institutions and people. Research is more and more global and relies increasingly on information technology. Astronomy is a prime example of this because astronomers are a very networked community; the research activities are based on the sharing of international infrastructure, the free flow of information and the need to move massive amounts of data around the globe.

Canada excels in this scientific field. It is ranked third in the world; it has an excellent publication record; it is regarded by the community as a partner of choice because of its strong national university-government network. The characteristics of this network are:

- Open communications with free flow of information and knowledge;
- The notion of community good;
- Access to facilities is guaranteed if you participate in the Structural Capital (building new facilities);
- Intellectual property is shared and not protected; and
- Relationships based on the scientific reputation of individual and mutual recognition.

Canadians have played outstanding roles in the development of modern astronomy which has derived downstream benefits. Canadian astronomers represent a numerically small, but scientifically highly respected community on the world stage. For example, Canada invested[9] $30M in 1999/2000 in the field of astronomy - the least per capita of the G-7 astrophysics countries' total investment of $3.4 billion - but ranks third in the world after the US and France for its contribution to astrophysics. In fact, in an international study looking at the contribution to science through a study of citation index, astronomy ranks the highest (38% over the world average) of all science field activities in Canada.[10]

Beyond the scientific record, Canadian firms have been very successful in generating benefits from international astrophysical facility construction and instrumentation. Benefits from astronomy and astrophysics also frequently arise through application of instrumentation advances developed in the course of the research in a number of other fields. This latter area has proven to be an exceptionally fruitful area for industrial and medical innovation, and there are many examples of such applications both from Canadian astronomy and from that in other countries.

An illustration of the economic potential lying within the links among astronomy, technology, and industry is the story of AGRA-Coast (Port Coquitlam, BC[11]). In the late 1970s, this firm gained a $4M contract to build the dome for the Canada France Hawaii Telescope (CFHT) on Mauna Kea, Hawaii. Their traditional road and bridge work benefited directly from their successful attack on construction problems at high altitude and freezing temperatures, and the direction of the company changed toward one of advanced technology that now includes design and simulation, satellite tracking, and sophisticated theme-park rides among other applications. The company is now the world's leading builder of enclosures for large telescopes: it has constructed the domes for the Keck I and II optical telescopes; the Japanese Subaru telescope; the Gemini North telescope (all on Mauna Kea); Gemini South; two UK optical telescopes on the Canary Islands; radio antennae for the Owens Valley (Caltech) mm array; and other US and Spanish telescopes.

The AGRA-Coast contracts for the Gemini project alone are valued at $44M, which should be compared with Canada's original $38M capital investment to enter Gemini (generated by combined contributions from NRC, NSERC, and the WESTAR Corporation). Its work on all telescopes since the 1970's has now grossed $150M. The overall effect is larger when it is realized that the firm is now a knowledge-based one with annual contracts above $10M/yr. AGRA-Coast has subcontracted many parts of its telescope enclosure work to smaller firms in the Vancouver area, stimulating the technology businesses there.

Smaller Gemini contracts have been let to a variety of other BC companies including Ramsay Engineering, Stevested Engineering, and ASA. About 80 Canadian firms expressed written interest in the Gemini work at preliminary stages, indicating the range of possible economic opportunities. The Herzberg Institute of Astrophysics (HIA) itself was awarded contracts worth about $9M for more sophisticated instrumentation.

An earlier concrete example of industrial spin-off came from the Toronto-based firm DSMA Atcon Ltd., which in the mid-1960s invested $0.25M in a design for a 4-m-class optical telescope. The design work enabled them later to bid successfully for numerous components of telescope projects in Italy, Germany, and Australia, as well as nuclear research and space robotics projects. The firm's leverage factor of its initial astronomy involvement was estimated by the company president to be 40 to 1.

The stimulation of economic activity generated from the construction and operation of major astronomical facilities is not restricted only to dome and telescope construction. The spin-offs reach very far into the structure of the economy. The consultant firm, The ARA Group undertook a small survey[12] of Canadian astronomy university departments that revealed a surprisingly high proportion of faculty members have some sort of "practical" interaction with industry and/or government. Recent concrete examples, of such interactions at Canadian universities, showed that the impacts can be very substantial:

- Two students from École Polytechnique started a small company, Matrox, in 1979 to produce electronic cards for storing numerical images at the Mt Mégantic Observatory (OMM). Today Matrox is a world leader and has annual sales of about $500M, and export sales in excess of $200M annually.

- In 1984, two researchers at OMM (the director and a computer science graduate) developed a digital system for computer display of astronomical images. The graduate subsequently founded the software company Softimage. This firm has grown into a world industry leader in professional visual content production, with tools for creating 3-D and 2-D animation, as well as for creating, editing and finishing video programs.

- The University of Alberta's Department of Electrical and Computer Engineering, in collaboration with HIA and the Alberta Microelectronic Centre (now the stand-alone Alberta Microelectronic Corporation), led to the development of prototype antenna/receiver modules using state-of-the-art superconducting-insulator-superconducting (SIS) technology. (This is of use in astronomy, but also industry for use in various mm and sub-mm engineering projects and other cryogenic SIS projects.)

- The University of Lethbridge, Alberta, is currently involved in a project to rapidly measure the amount of water vapor in the atmosphere for phase correction of radio interferometer data. This may lead to a method of rapidly determining air density variations at airport runways, with the potential of serving as a powerful diagnostic tool for airline pilots. A second example is the gold-coated nickel metal meshes which were developed for a spectrometer launched on the Infrared Space Observatory (a European Space Agency mission). Later these meshes were found to be ideal surfaces for rapidly growing skin for severe burn victims. The meshes have been used at major US burns institutes.

- The Institut National d'Optique (INO, Ste Foy, Quebec) carries out extensive research and development in photonics, optics, electro-optics, and lasers (annual operations $18M). It has performed extensive astronomy-related work for CSA, European Space Agency, and other contractors. It employs several astronomy-trained staff. Its expertise matches many areas of the Long Term Plan.

- At the University of Victoria, BC, recent research on the time ephemeris of the earth has refined (by roughly two orders of magnitude) the relationship between earth-based and solar-system based clocks. This has implications for solar system spacecraft ranging.

- Queens University, ON, has developed algebraic computing methods that have been used in the financial industry, and it has provided computing consulting to the petroleum industry.

- The University of Waterloo's astronomical imaging techniques have been used in medical imaging and remote earth sensing research.

- York University, ON, established a spin-off company, Stellar Optics Research International Corporation, which markets products and consulting in the area of black and white materials and surfaces. This company got its start as a result of two contracts: one related to providing advice on materials or surfaces suitable for flat field astronomy CCD images, and the other for advice on black surfaces suitable for the baffle of the Far-Ultraviolet Spectroscopic Explorer (FUSE). It now supplies flat fielding screens to astronomers worldwide. In addition, York astronomers have learned how to make ultra-precise spheres, which will be used for the new international density standard, and astronomical software has been transferred to the private sector for applications in research and education.

- At the University of Toronto's Canadian Institute for Theoretical Astrophysics (CITA), ON, faculty have been involved in developing microwave detectors for use on satellites. It has earned income providing images and animations of galaxy mergers to the entertainment industry.

- Canada's enormously successful RADARSAT project employed cleverly implemented, aperture synthesis for coherent signals, a technique developed by radio astronomers for imaging the sky.

NRC has also contributed via its Herzberg Institute of Astrophysics by providing industrial assistance and R&D contracts. Among many commercially related activities in the past three years, the Institute has:

- Worked with the Centre for Research in Earth and Space Technology (CRESTech, one of the Ontario Centres of Excellence), CSA, and the federal Natural Resources Canada, developed a version of a data acquisition and recording system in the so-called S2 format for correlating Very Long Baseline Interferometry (VLBI) data from radio telescopes. CRESTech has made sales of approximately five dozen S2 recorder, formatter, and playback systems to about a dozen countries including the USA, Japan, Australia, India, and France with revenues of about $6M to date, equal to the S2 development costs. Sales are continuing, and association with the HIA data correlator facility, which is one of the four most advanced in the world, is an added strength.

- Assisted a small BC company (Globus Celluar Inc.) in developing a cell phone antenna that is more energy-efficient and directs less radio emission into the user's head; if successful, the potential market is very large indeed.

- Developed a new lens coating technique, and have begun discussions with businesses for its widespread commercial potential.

- Built and made available a "virtual antenna" test facility to private industry.

- Worked closely with COMDEV (Cambridge, ON) on several technical support issues for satellite instrumentation, notably the fine error sensors for the recently launched FUSE spacecraft supported in part by CSA with NASA.

- Provided technical assistance to Bristol Aerospace on its LIDAR (space) telescope project, and worked closely with other firms developing space instrumentation including EMS-T (formerly CAL, Ottawa), Routes Inc., and MPB Technologies (Montreal).

- Assisted AGRA-Coast on mm-wavelength antenna technical knowledge necessary for them to bid on the international ALMA project.

- Has been actively pursuing a business development process to license data archiving and distribution software that can be used in library and commercial data management operations.

Evidence of impact of astronomy activities can also be found in the international literature. The American decadal review of astronomy and astrophysics published by the National Academy of Sciences in 1991 lists at least a dozen other major applications of astronomical techniques in various areas of industry. These include image processing software in automobile and aircraft labs; communication antenna testing; low-noise components for the communications industry; development of highly sensitive photographic films by Kodak; infrared-sensitive films for aerial reconnaissance; and X-ray baggage scanners for airports.

The 1999 US decadal review is in preparation, and demonstrates an accelerated pace of exploitation of astrophysical instrumentation. Some examples of recent practical applications driven by astronomy and astrophysics include:

- Development of ophthalmic instruments which can image the retina of an eye and measure an individual's eye aberrations.

- Better X-ray equipment (lower dosages, increased sensitivity, lower sample or tissue damage, ability to use single photons).

- Better imaging equipment at UV and optical wavelengths: applications include industrial and tumor imaging.

- Better infrared imaging equipment for night vision), semiconductors and medical detectors.

- Better communications at radio wavelengths (lower noise and interference, wider bandwidth), aircraft landing systems.

- Better spectroscopy: applications include trace chemical analysis, (e.g. pollution, global warming, ozone depletion).

- Data and image analysis; applications for underwater imaging, mapping atmospheric aerosols, breast cancer detection, genetic decoding, and police work.

- Precision timing: applications include satellite communications and positioning, tracking cellular phone calls (including 911 calls), geodetics, GPS handheld transmitters.

- Data analysis and computation; applications include: FORTH computer language (commercialized from NRAO work and now used in many sectors, including Federal Express delivery and tracking systems, automobile engine analyzers, etc.).

It has proven difficult for economists to put a "hard number" on the economics of fundamental research advances in any field. This is due to a variety of reasons, in particular the lengthy delay required between the time that research is done and the time application occurs, a delay often measured in years and decades. Another reason is that applications are hard to track, since they frequently arise in areas completely unexpected and unrelated to the field in which the research was originally done (as is probably obvious from some of the examples above) and are thus hard to track. Furthermore, before deciding on the best estimate of economic impacts, we must first address the related issues of incrementality and attribution. On incrementality, some industrial impacts may have occurred even if the government research investment had not been made. For example, if it were not for the success of one project; AGRA-Coast, the results for Canada would have been much different. On attribution, many industrial impacts result from the cumulative effect of many research projects in many fields carried out over many years, and tracking such benefits is a challenge. Thus counting only the most recent research expenditures, or if only a few large examples, or only those in one particular field, undoubtedly either underestimates the true costs or the true returns on investments.

9. INDUSTRIAL RESEARCH (MATERIALS)[13]

NRC's Industrial Materials Institute (IMI) has positioned itself as a central force in materials processing at the national and international level, by pioneering at NRC a visionary model of industrial collaboration that has had an important impact on Canada's manufacturing innovation process.

1. Astrophysics Network

Key to Technology Transfer

- Open exchange by researchers
- Recognition of Canadian excellence
    - Astrophysics had the highest citation impact record, at 38, for 22 disciplines in papers from 90 countries
    - Highest ranked citations of 13 optical telescopes (4 mm)
- Canadian equipment manufacturers brought in early on

IMI was motivated by a desire to enhance its relevance to industry by including multi-source input into Process Modeling and Optimization programs at the earliest possible stages. It paid close attention to the phenomena of the rise and disappearance of consortia of the 1980s and developed from the experience a simple, elegant model for cooperation and networking focused on manufacturing: the Special Interest Groups (SIG).

The SIG differs from a consortium in a few important ways. First and foremost, SIG activity is restricted to pre-competitive research and entry fee is relatively low (only $30,000 annually for the largest companies and $10,000 for smaller players). Because of the relatively low investment it reduces the need for a fast return on the investment. Finally, each member of a given SIG signs on directly and only with NRC rather than with each other, leaving IMI in control of the work it performs in response to membership input.

With its focus exclusively upon pre-competitive research, IMI's SIG model occupies the "middle ground" of the research process, falling between short-term market driven projects and long term more fundamental research. It is research that benefits all members without jeopardizing the intellectual property concerns of industry or their competitive edge, because each of the groups is focused on improving the fundamental processes involved in the making of parts. Furthermore, better understanding of the behavior of materials during molding, casting and forming, improves drastically the potential for efficiencies across the board. Raw materials providers and equipment makers see this value and are eager to participate because of the synergy achieved amongst the manufacturers themselves.

Each SIG targets a specific branch of manufacturing but is structured according to the same template, giving stability and simplicity to the concept. Members meet on three occasions annually to discuss and prioritize the research initiatives that IMI staff is pursuing as a result of Group input. Each member has one vote and can send only a specified number of representatives so that the big players can't dominate the smaller ones. There must be at least 60% support from membership for an initiative to be pursued, ensuring that no narrow agendas dominate and that the research being done is relevant to the needs of the collectivity.

The collaborative and fertile environment created by the SIGs leads, of course, to new ideas for products that individual members wish to bring to market which cannot be done within the boundaries of a SIG. IMI can pursue these opportunities with individual participants through a Special Project arrangement made with the company in question. IMI earns more than $3 million each year from these Special Projects as well as an equivalent amount from in-kind benefits such as equipment and access to facilities.

IMI has been directly involved in the creation of seven SIGs, in the fields of injection molding, blow molding, die casting and thermoforming, surface technology, polymer blending and polymer foam technology. The impact these initiatives have had upon the Institute and the Canadian manufacturing sector in general has been important. From an NRC financial perspective, IMI receives about $1 million annually in Group fees in addition to the above-mentioned $3 million generated by Special Project. This amount represents about 30% of the Institute's budget, a figure that matches the percentage spent on SIG driven research. It is, therefore, a largely self-funding endeavor that, in fact, helps rather than hinders the long-term exploratory work to which IMI continues to direct a large portion of its resources.

Benefits enjoyed by IMI staff working within the SIG programs have been significant in other ways. Individual researchers are rewarded with the knowledge that they are producing results which improve the competitiveness of Canadian industry and which solve the problems of today for better performance tomorrow. They are an integrated part of a dynamic multi-billion dollar whole, whose expertise is highly valued and sought after. There is also the opportunity to access equipment and facilities, paid for by group members, which would otherwise be beyond the means of public funding.

For example, initially the injection molding initiative attracted 12 members at its launch; there are currently 130 companies involved in the SIGs as well as several universities. This enthusiasm on the part of concerns such as Bombardier, Husky, Noranda, Dow and the Big Three auto makers is due to the fundamental importance of improving parts quality and reducing cycle rates. IMI's expertise in Process Modeling Optimization and Simulation provided solutions to these imperatives that are as important to construction, electronics and packaging as they are to the aerospace and automotive industries. IMI targeted specific sectors for SIG establishment precisely because each is crucial to the basics of manufacturing - make better parts faster and cheaper.

Because of the pre-competitive, cooperative nature of the SIG formula companies have entered into no fewer than 150 Special Project arrangements with IMI. In one example, IMI expertise helped General Motors reduce its rejection rate for a major component from 40% down to 3%. In another, Twin Pack, the largest plastic bottle producer in Canada saw a reduction of 0.7g from a 33g blow molded pop bottle. When that 0.7g is multiplied by the 28 to 30 million units produced daily, the savings are considerable. Special Projects also helped Chrysler optimize the making of piston heads, Pratt and Whitney a $2.3M element auxiliary gearbox and Volkswagen a sophisticated wheel to mention just a few others. Companies join the SIGs because they get results, they benefit from new knowledge, ideas and capabilities shared within the Groups as well as from the Special Projects that focus upon particular, proprietary concerns.

The success of IMI's SIG model has also had an impact on employment. To be member, a company must have manufacturing facilities in Canada, and all R&D stemming from the Groups is conducted within Canada. This has encouraged companies eager to reap the benefits of membership to invest in Canada rather than abroad. IMI estimates that 150 new Canadian jobs can be directly attributed to the existence of the program. Graduate students from participating universities that are involved in SIG activity also find the Groups to be employment generators. Many of them are hired to premium jobs by companies that otherwise would not have known they existed.

2. Sector Interest Groups (SIGs)

Key to Technology Transfer

- Closed groups with periodic meetings and open exchanges within the group
- Variable fees structure to accommodate SMEs
- Shared vision on research needs
- Industrial firms decide how they use the information developed in the group, or what actions follow-on they want to take.

10. INDUSTRIAL SUPPORT

A third example is the IRAP program, whose objective is to support SMEs across Canada in their innovation process. The network consists of over 260 fulltime Industrial Technology Advisors (ITA) and the Canadian Technology Network (CTN) with over 1000 additional part time advisors.

Basically, the ITAs act as knowledge brokers and facilitate linkages to ensure that SMEs have timely access to the knowledge they need in their process of innovation. The program also provides seed financing (1/3 to 1/2 of the cost on the average) to the SMEs towards part of their R&D activities, and in return for which they agree to exploit the results of the research in Canada for a period of at least five years. Because of the proprietary nature of the projects, in this case, there is a strict respect for confidentiality to understand the nature of the issues around the technical opportunity or problem so that the ITA and his / her network can provide advice and appropriate linkages to specialized expertise and facilities. One of the most important skills of an ITA is to build mutual trust in managing the network relationships. The characteristics of the IRAP-CTN network are as follows:

- Highly qualified people that provide credibility in the exchange of knowledge and the building of relationships both with the firm and the S&T community;
- The financial assistance is tied to innovation and shared risk;
- The network is a people network assisted by information technology; and
- Relationships are based on trust and the pursuit of common objectives.

In spite of a number of prior IRAP benefit studies, it has still been hard to quantify the direct impact of IRAP support, i.e. the incremental contribution. It is well recognized that IRAP firms generally perform well. But it is not well understood the extent to which the growth of IRAP supported firms is directly attributable to the IRAP / ITA process of advice and financial assistance. Statistics Canada's[14] comparative study indicated that IRAP funding produced higher growth in value added and in exports, but not necessarily in jobs growth.

NRC's Slater Grimley undertook another approach. In the early 1980s, they tried to assess the difference in performance of 6 matched pairs[15] of firms with and without IRAP funding. Even these pairs were found to have too many differences to show significant attributable results to IRAP assistance. Other IRAP evaluations of sales and job creation have included assessments of incrementality, which generally show some positive indications of IRAP support. Joel Baum of the University of Toronto in a study (1999) studied failures in 800 companies, broadly defined as biotechnology firms; he found that job growth of IRAP supported firms was slightly higher than for non-IRAP funded bio firms.[16]

Slater Grimley used 1995 data to generate a report on IRAP job creation, which dealt with all IRAP projects ending in 1991-92. He concluded that the government returns via the extra tax revenue generated are very large. Even if the government must borrow the money for IRAP at an interest cost of 10% per annum, its annual expenditure is recovered with interest within the next five years, plus an added government profit of 10-25% per annum. IRAP achieves this remarkable result by helping remove the technological uncertainties so firms' ventures may proceed to commercialization with little or no further public assistance. This special "leverage" by IRAP produces very high attributable tax returns for the government, helped by IRAP's spreading of the risks over a large number of projects in all industry sectors. With today's interest rates the government's returns from its IRAP expenditures would be even greater.

Firms with IRAP support enable them to undertake more advanced R&D projects than they would do alone. Firms without IRAP would undertake lower risk or shorter-term projects as they could get financing for them, but most likely with lesser job creation or job saving results. Some firms would disappear. Statistics have shown that the failure rate of IRAP firms over five years is 20% below that of non-IRAP firms. The returns to the financial sources of the firms in question would also tend to be lower without IRAP, again producing lesser economic gain. Some would be lucky enough to do quite well.

It is one of the outstanding features of IRAP that it utilizes the economic system in such a way as to give satisfactory returns on the investments of both the general taxpayer and the private client for relatively little government expense. Grimley's report shows an average of $150,000 sales per IRAP job for the early 1990s period. For the 8,740 jobs due to IRAP, this gives $1.3 billion in direct sales per annum. This is a ratio of attributable sales to IRAP expenditure of 17:1. The results of long-range multi-year IRAP projects were to follow after the study phase and so are not fully reflected in the data obtained for this study.

More recently, IRAP[17] assessed the performance of university spin-off firms in Canada with data from other key countries and the difference in downstream socio economic impact of IRAP versus non-IRAP supported university spin off firms in Canada. In the past it has been hard to find a large enough sample of matched firms with and without IRAP funding. The study covers 344 IRAP supported firms (42%) out of 816 university spin-off firms identified to July 2001, but excluding firms created in 2000 or 2001. *University spin-off firms now make a major socio-economic contribution in Canada*[18] *– these are estimated to have involved more than 19,000 jobs and $3.4 billion in sales in 2000.*

Creation of university spin-off firms is just one method of creating socio-economic benefits from the university system. The prime contributor is from the highly qualified personnel (HQP) that contribute to innovation with increased salaries upon graduation and the downstream taxes and follow-on wealth multipliers etc. Other major contributions are technology transfer or expertise to established firms - either by licensing or consulting. However, the socio-economic benefits from spin offs may be more readily measured, and have been rising rapidly over the past 5 years. The study shows that the spin-offs are mostly in the new economy. University technologies in the more traditional areas such as engineering, electrical and chemical sectors, tended to be transferred directly to established firms in Canada or abroad instead of via creation of spin-off firms.

For a firm to be classified as having received IRAP support it would have received at least $15,000 or more beginning within the first 5 years of formation of the firm, i.e. during its early stage after incorporation, and not, as in some excluded cases which received its first IRAP support say 6-10 years after incorporation. The 816 spin-off firms in the study were created between 1921 until mid 1999. The rate of formation of spin-off firms increased from 8 per year in the 1970s to 42 firms per year in the 1990s; and up to 70 firms in 2000. Fifty three percent (53%) were formed from 1992-1999.

The study assessed a number of parameters such as closure and/or bankruptcy ratio, sectorial distribution (albeit mostly high technology area) and their contribution to jobs and sales.

From a macro perspective, the failure rate of Non-IRAP firms was approximately the same (46% IRAP and 54% Non-IRAP). This came as a bit of a surprise but suggest that the IRAP firms are likely to be a representative sample of the total because the distribution by sector is the same: life sciences (39% IRAP and 61% Non-IRAP); electronics and communications (51% IRAP and 49% Non-IRAP); software (44% IRAP and 56% Non-IRAP), and others (42% IRAP and 58% Non-IRAP). In 1999, IRAP supported spin offs seems produced a slightly higher number of jobs 56% but only 48% of the sales.

Where the IRAP supported spin offs really distance the Non-IRAP supported firms is when they go public. So far out of the 816 firms, 58 (7%) have gone public of which 6 were taken over and one was closed. It is estimated that the 51 firms as of July 2001 created 10,500 jobs and $2.5 billion worth of sales. The data shows that these IRAP supported firms predominate in numbers of firms (58%), much higher sales (80%) and higher number of jobs (77%). IRAP-supported firms also to get to the public market exchanges faster than the other firms after they are incorporated either via IPO or reverse take over.

Independent assessments of performance again show that IRAP assisted firms perform better as seen in various national indicators such as the *Globe and Mail* Top 1000 performers (92%) or *Profit Magazine's* Top 100 Technology companies.

3. IRAP linkage with SMEs / Spin Off Firms

Key to Technology Transfer

- Open research at university leads to technology transfer to ongoing firms via consulting or licenses, or to creation of new spin off firms.
- IRAP ITA provides SMEs with one-on-one:
  - confidential advisory service and linkages to universities and other sources of expertise and facilities (e.g. NRC), and
  - where necessary, funding for specific R&D projects.
- Each ITA has an extensive personal network and links to other networks such as the 1,000 members of the Canadian Technology Network.

11. CONCLUSIONS

Three quite different modes of innovation networks have been discussed and yet each one provides significant economic benefits. But they all involve important human interactions:

- Are a human community capable of providing diverse meaning to information with the result of creating new or improved knowledge;

- Provide multiple perspectives on issues and alignment in a dynamically changing environment;

- Encourage diverse viewpoints that need deeper analysis. People with differing backgrounds and expertise can provide a much broader focus, essential to grasp the essence of the core issues;

- Enhance greater proactive involvement of human imagination and creativity;

- Harness the tacit knowledge;

Therefore active use of key network nodes by collaborating users is the key to user value from networks.

Chapter Seven
R&D ALLIANCES AND NETWORKS INDICATORS AT THE DIVISION OF SCIENCE RESOURCES STATISTICS, NATIONAL SCIENCE FOUNDATION

Francisco Moris and John Jankowski
Research & Development Statistics Program (RDS)
National Science Foundation

1. INTRODUCTION

Alliances, joint ventures, and other networking activities allow firms to share research and development (R&D) costs, pool risks, and enjoy access to partners' firm-specific know-how and commercialization resources. Increasingly, corporate R&D requires a combination of technology *exchange* and technology *co-development* through strategic alliances and other linkages in order to excel in innovation and market performance. At the same time, collaborative networks are not without risks. For example, unintended transfer of intellectual property and competitive implications of collaborations have occupied business and public policy making and research for a number of years.

The Division of Science Resources Statistics (SRS) of the U.S. National Science Foundation (NSF) has been interested in the role of *strategic research partnerships* in the science and technology enterprise, especially over the last decade. SRS has funded survey activity, database development, research, and workshops on the topic. It has also tracked trends and disseminated advances in the understanding of this phenomenon in its analytical outputs. This chapter discusses these activities and concludes with a summary of major data needs from a public policy perspective.

2. S&E INDICATORS: R&D PARTNERSHIPS AND TECHNOLOGY TRANSFER

The biennial NSF publication, Science and Engineering (S&E) Indicators, has analyzed indicators related to cooperative research and technology transfer across the major R&D performing sectors – industry, universities, and government agencies or laboratories – since 1972. Text Table 1 shows a summary of these indicators listed by the year in which they were first included in the S&E report.

There was a proliferation of alliance-related indicators sponsored or used by SRS/NSF during the 1990s, stemming, in part, from reporting requirements of certain Federal statutes. Examples include data on joint research activities by private companies under the provisions of the National Cooperative Research and Production Act of 1993 (Industrial joint research ventures) and data on Federal laboratory technology transfer after the Federal Technology Transfer Act of 1986 (Federal technology transfer indicators). Similarly, activity and data on university patenting also increased markedly after limitations on patents rights related to Federally-funded research were lifted by the Bayh-Dole Act of 1980. Still, other indicators were the result of NSF initiatives in the form of funded studies or database development (International strategic alliances, University-industry research centers, and Patent/article citation indicators).

Text Table 1

R&D expenditure flows across sectors (1972)
Bibliometric co-authorship indicators (1980)
University patents (1982)

Federal technology transfer indicators (1991)
Industrial joint research ventures (1991)
R&D expenditure flows across countries (1991)

International strategic alliances (1993)
University-industry research centers (1993)
Foreign location of R&D facilities (1993)

Federal cooperative technology program dollar support (1996)
University technology transfer indicators (1996)
Patent/article citation indicators (1996)
Defense dual use programs (1996, 1998)
Advanced Technology Program data (1998)

3. NSF-FUNDED DATABASES

3.1. CORE and NCRA-RJV Databases

3.1.1. Overview

CORE (Cooperative Research) and NCRA-RJV (National Cooperative Research Act – Research Joint Ventures) are NSF-sponsored databases on U.S. research joint ventures (RJVs). The National Cooperative Research and Production Act (NCRPA)[1] of 1993 allowed research – and related development and production activity – to be performed in collaboration with industry partners. However, companies forming RJVs need to notify the U.S. Attorney General (Department of Justice) and the Federal Trade Commission in order to enjoy the benefits of the NCRPA statutes.[2] The notifications are published in the Federal Register. CORE is maintained by Al Link (University of North Carolina-Greensboro), and NCRA-RJV by Nicolas Vonortas (George Washington University).

The databases differ in their basis for classifying data: CORE in terms of joint venture units, NCRA-RJV in terms of individual members. Both are available from 1985 forward (see Figure 1).[3] They each have been used extensively in NSF's Science and Engineering Indicators since 1991 and in numerous research articles based on NSF-sponsored and other academic research.

Figure 1. Domestic Strategic Research Partnership: 1985-2000

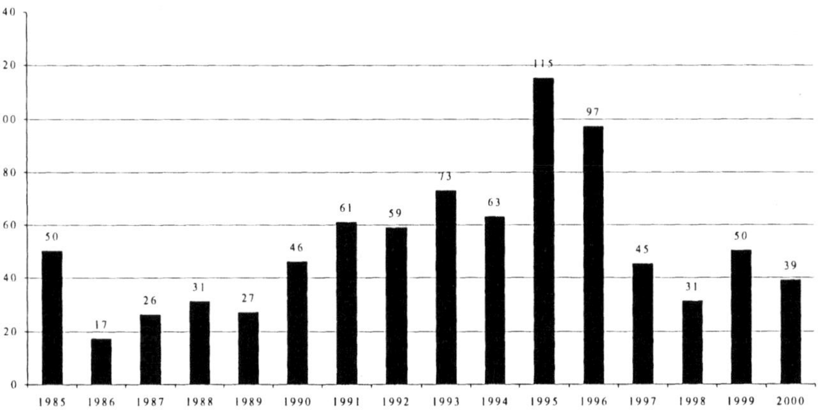

Note: Data are annual counts of new domesitc strategic research partnerships under the US National Cooperative Research Production Act.
Source: CORE and NCRA-RJV databases

3.1.2. Data Items

1) Number of new RJV filings
2) Classification of RJVs by industry and technology area
3) Number and industry of venture members
4) Participants identity and characteristics: company name, public/private status, industrial classification; university; government agency or laboratory; entry/exit dates
5) Research focus: process vs. product; environmental, safety research
6) Information on plans to patent, license, or trademark
7) Additional data available for public companies (obtained from other sources): sales, R&D investment, physical capital investment, employment, primary and secondary line of business by 4-digit Standard Industrial Classification (SIC) code

Timeframe: 1985 – 2000 (latest update)

3.1.3. Advantages

- Publicly available multi-industry time series of research joint ventures in the U.S.
- Based on company records, collected according to statutory definitions.[4] (However, administrative data have their own limitations, as discussed next.)

3.1.4. Limitations

1) Quality as a RJV Indicator

The filing counts, at best, represent the tip of co-operative activities among firms. Informal co-operation is not captured. Even for coverage of formal co-operative activities, there are shortcomings since the RJV counts are not obtained by a dedicated survey. While data based on survey responses have their own statistical caveats, counts based on administrative/regulatory filings are affected by the fact that the decision to file is a separate strategic and legal decision. Thus, filing counts may not reflect RJV activity accurately, possibly understating or overstating actual activity. For example, the counts might be biased downward whenever the overall antitrust climate in the U.S. is viewed as relaxed. Alternatively, a RJV sponsored by a trade association may register an inordinate roster of members interested in an umbrella protection, even if actual innovative activity by some of its members is minimal.

2) Non-R&D Activities

Neither database keeps track of whether the RJV has other complementary objectives such as cooperative production, as allowed by the 1994 NCRPA. It is our understanding that some indication of production intent is in the filings. However, the databases do not record this information.

3.2. CATI-MERIT/NSF

The Cooperative Agreements and Technology Indicators (CATI) database consists of counts of international technology alliances (see Figure 2).[5] It is compiled by John Hagedoorn at the Maastricht Economic Research Institute on Innovation and Technology (MERIT), the Netherlands, in part with NSF funding. The database is based on announcements in newspapers, trade publications, and other public sources. It excludes manufacturing or marketing alliances and it is tabulated according to *country of ownership* of the firms forming the alliances, as opposed to, say, the location of the alliance.

Figure 2. International Strategic Research Partnerships: 1980-2000

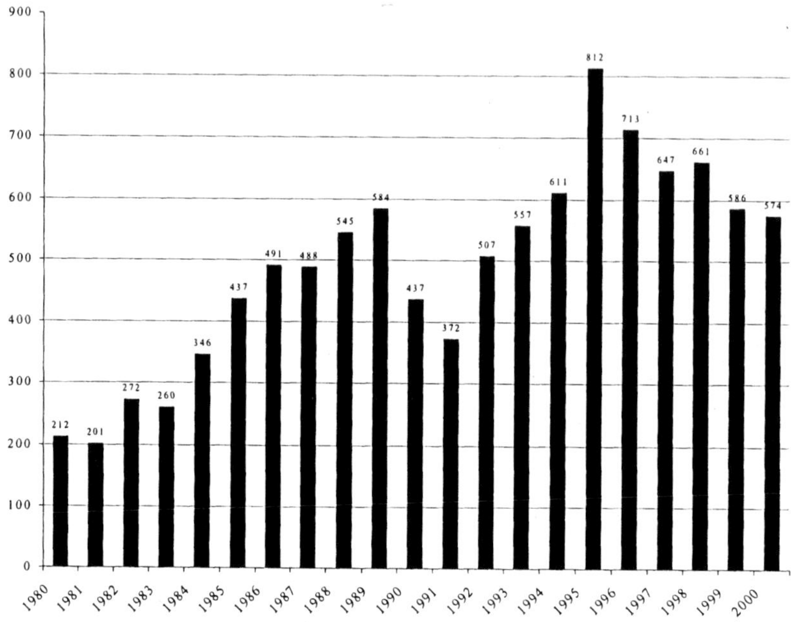

Note: Data are annual counts of new international strategic research partnerships.

Source: CATI-MERIT database.

3.2.1. Data Items

Counts of international technology alliances, classified by country of ownership and technology. Timeframe: 1960 – 2000 (latest update)

3.2.2. Advantages

Time series of RJVs with systematic international character. CORE and NCRA-RJV contain some information on participation of foreign-owned members, but only for U.S.-based joint ventures.

3.2.3. Limitations

1) Coverage

Because data are limited to activities publicized by the firm, agreements involving small firms and certain technology fields are likely to be underrepresented. Another limitation is that the database draws primarily from English-language materials.

2) Classification Criteria

i. Ownership vs. location of alliances. Both classification criteria are of interest since they provide different information. CATI-MERIT uses the ownership criterion, which introduces double counting issues for alliances involving parents/subsidiaries of more than one country. The double counting is taken into account in the various tabulations. However, difficulties still remain since country of ownership varies along the corporate structure. Assumptions on how far up the ownership chain one should go and how to treat minority holdings, for example, may affect the usefulness of some country-specific counts based on ownership criteria.

ii. Industry vs. Technology. CATI-MERIT is classified according to technology activity of the alliance. Industrial classification of alliances can facilitate linkages with other economic databases.

3) Plain CATI counts of research joint ventures are not weighted by the value of the agreement or by other appropriate metrics (Copper and Merrill, 1998).

3.3. IRI/CIMS R&D Innovation/ Large Firm Performance Database

NSF co-funded the Industrial Research Institute/ Center for Innovation Management Studies[6] (IRI/CIMS) survey on sources and allocation of R&D spending of IRI company members for fiscal years 1992-1998 (SRS 98-13597). The survey questionnaire is organized in terms firm, business segment, and laboratory level data (Bean, *et al.*, 2000).

3.3.1. Selected Data Items:

- Basic and applied R&D spending; process development and product development spending
- Support services
- R&D personnel
- Sales and costs savings during year of survey 'attributed to' new or improved products, services, and processes commercialized during 1993-97.

3.3.2. Advantages

- Data by firm, business segments, and corporate laboratory level
- Additional spending detail: technical R&D vs. support services.

3.3.3. Limitations

- Little or no representation of small firms
- Minimal information on innovation costs, other than R&D.

4. RELATED NSF SURVEY ACTIVITY

There is no dedicated survey on R&D alliances and networks at SRS/NSF. However, new survey items related to this collaborative activity have been introduced in existing surveys of R&D performers (universities, industry) as well as in other survey activities across the Division.

4.1. NSF Industry R&D Survey

Starting with 2001 data, detail of extramural R&D (R&D performed outside the company, but within the U.S.) will be available by type of contract organization: for-profit, university, non-profit organizations. Previously, only total extramural R&D spending was requested.

4.2. NSF Survey of Academic R&D Expenditures

Pass-through R&D funding: total dollar amounts sent to (starting with 1996 survey) and received from (starting with 2000 survey) higher education and all other institutions.

4.3. NSF's IT Innovation Survey

This one-time survey, in progress as of early 2002, is designed to determine how businesses plan for and produce innovation in Information Technology (IT). The survey targets both producers and users of IT. It includes two questions on activities related to R&D alliances:

- Proportion of firm's alliances with other firms, universities, or government agencies involve R&D activities (none, less than 25 percent, 25-50 percent, more than 50 percent)
- Proportion of firm's annual R&D expenditures directed toward alliances, partnerships, or joint ventures (excluding contract R&D activities).

4.4. Bibliometrics and Patent Database

Available data related to cooperative R&D outputs include co-authorship, university patents, patents granted to foreign investors, and international patents.

4.5. Survey of Doctorate Recipients

Data on overseas communications, visits, and research activities by U.S.-based doctorate recipients (1995 survey only).

5. NSF-FUNDED (OR CO-FUNDED) STUDIES AND RESEARCH GRANTS

5.1. Private Strategic Research Partnerships

Data from NSF-funded alliance databases have been used extensively in academic papers, some based on NSF-funded studies, as well as in NSF's S&E Indicators.[7] Studies on special topics such as intellectual property protection in alliances and licensing agreements (another form of technological linkages) have been funded.

5.1.1. Funded Studies on RJV Indicators

Project	Completion Date/Grant No.	Principal Investigator(s)
Research & Innovation Statistics	1999/E 01804X	National Research Council

This study, performed by the National Academies of Sciences in the late 1990s, identified major changes in innovative activities in the U.S. (Copper and Merrill 1998).[8] Trends discussed included changes in composition and orientation of R&D activities (service sector R&D, services R&D in manufacturing), changes in organization of innovative activities (alliances, technology acquisition, central vs. divisional R&D labs activity), issues in labor/career mobility, and linking of innovation indicators to relevant economic databases.

Project	Completion Date/Grant No.	Principal Investigator(s)
Network Indicators	Ongoing/0097948	Vonortas

This study aims to identify innovation network indicators by examining network formation, structure, activities, and research outputs (e.g., patents). In addition, the project will carry out case studies of individual networks based on company interviews.

5.1.2. Funded Studies on Special Topics

Project	Completion Date/Grant No.	Principal Investigator(s)
IP Protection and Research Partnerships	2001/SDEST 99-06140	Vonortas/Hertzfeld

NSF grant for a one-time survey of 250 firms (with at least one RJV in the NCRA-RJV database during 1995-98 period) on their strategies for intellectual property (IP) protection within research alliances.

Some findings (Hertzfeld, Link, and Vonortas 2001):

1) Patents are the most frequently used intellectual protection mechanisms to protect both existing (pre-alliance) IP as well as knowledge or technologies forthcoming from the cooperative R&D activity.
2) Existing patents are useful when negotiating new research joint ventures.
3) IP negotiations are more difficult in horizontal RJVs (collaboration between firms in the same industry) and when an alliance involves university partners.

4) Trade secrets protected by non-disclosure agreements, for example, were also used extensively as an IP protection mechanism.

Project	Completion Date/Grant No.	Principal Investigator(s)
Licensing Agreements	2001/SRS 96-15976	Link/Scott

Research grant to study licensing agreements in the chemical industry using existing databases (articles in business periodicals over 1993-97) (Link and Scott 2001). The study used a probit model to explain probability of *occurrence* of licensing agreement (not frequency of licensing agreements, which would have required a count model or a multinomial probit model).

Some findings (Link and Scott 2001):

1) Cross citations of patents: more licensing activity when licensee/licensor pairs cite each other.
2) Science linkage (based on counts of scientific references in a patent): probability of licensing agreement increases with science linkage of potential *licensor*; *decrease* with science linkage of potential *licensee*.
3) Product diversification: diversified pairs have higher estimated probability of reaching a licensing agreement.

5.2. Public-Private Strategic Research Partnerships

5.2.1. University-Firm Collaboration

Project	Completion Date/Grant No.	Principal Investigator(s)
University-Firm Linkages	1995/SBR 9305591	Rahm

NSF grant for a one-time survey of U.S. research universities on the characteristics of successful university-industry interaction, barriers to interactions, and costs and benefits (Rahm 1995).[9]

Project	Completion Date/Grant No.	Principal Investigator(s)
UIRCs Activities/ performance	1995/SBR 94-21958	Cohen/Florida

NSF grant for a one time-survey of the extent, characteristics, and activities of U.S. University-Industry Research Centers (UIRCs). For the purpose of the study, they were defined as university-affiliated research centers, laboratories, or other organizations that conducted R&D in science and engineering with industrial funds. These centers also received funds and other support from the Federal government (including, notably, NSF) as well as from state and local governments. Other activities of these centers include education and training (Cohen, Florida and Goe, 1994).

5.2.2. Science Parks

Science parks are business and research facilities typically close to and linked to a university or other research organization, often with state/local government support. Objectives include technology transfer and collaborative R&D between universities and young/small firms; creation and growth of high-tech firms; and student training.

The property-based aspects of science parks and incubators[10] differentiate them from other technical assistance programs that focus more on services or on time-use privileges of equipment or specialized facilities, for example.

Project	Completion Date/Grant No.	Principal Investigator(s)
Science Parks – Exploratory Analysis	2002/SRS 99-10785	Link/Scott

This NSF grant supported a study on science parks indicators, growth factors, and the impact of science parks in university academic missions and technology transfer activities. Link and Scott (2001), one report based on this study, formulates an econometric model to explain inter-park differences in employment growth rates in the U.S. The authors found that a shorter distance to university campuses, the presence of private venture capital, and having business/industrial sites for sale and lease (as opposed to for lease only) had a positive effect in science parks employment growth. These effects were estimated for different technologies and U.S. regions.

In another report based on this grant, the researchers found that science parks have a positive effect on universities' research output (Link and Scott 2002). In particular, universities with a formal relationship with a science park realize greater benefits in terms of research output, as measured by publication and patenting activity, after controlling for total university R&D budget. Other significant factors in the relationship between science park activity and university research output includes the age of the science park and the percentage of faculty engaged in university/science park activities.

5.2.3. CRADAs

Project	Completion Date/Grant No.	Principal Investigator(s)
Interviews on CRADAS Metrics	1999/E08885X	SRI International

Funding provided to assist the U.S. Department of Commerce in an SRI-conducted survey on attitudes towards and metrics on Federal Government Co-operative R&D Agreements (CRADAS). Data are used in S&E Indicators.[11]

6. CONCLUSION: DATA NEEDS ON RESEARCH JOINT VENTURES

Given the limitations associated with a separate national survey on R&D alliances, both at the agency level and in terms of respondent burden, the strategy of SRS/NSF has been based on the expanded use of existing surveys of R&D performers. This approach can be complemented with the integration of S&T data with existing publicly-available economic databases, as described below.

Required Data (and Statistical Strategy):

1) Alliance **structure** (improvements in NCRA-RJV/CORE, CATI-MERIT, or other RJV databases):

- Members profile: corporate structure (subsidiary of MNC, small firm)
- Location data by country of alliance activity
- Equity vs. contractual alliances
- Financing structure and value of deals
- Termination/ duration
- Frequency data by firm: activity over time, repeated partners (see item 3 below)

2) Alliance **activities** (improvements in NCRA-RJV/CORE, CATI-MERIT, or other RJV databases):

- Non-R&D objectives - sourcing, production, marketing
- Other technology linkages – licensing, other technology exchanges
- Crosswalk of technologies in CATI-MERIT with NAICS, ISIC Rev. 3.
- Industry of participants (useful to track multi-industry and vertical alliances)

3) **Longitudinal data** by firms and/or establishment (data mining of existing RJV databases):

Longitudinal data on alliances structure and activities (see items 1 and 2) would allow tracking changes in R&D organizational strategies over the life of a firm or set of firms, as suggested by the life cycle approach to networking and other technological linkages (Auster 1992, Oliver 1994).

This approach integrates the evolutionary, transactions cost, and resource-based views of the firm.[12]

4) Use of **human resources data** (e.g., labor, education statistics) to track microeconomic or social impacts of innovation activities of alliances (database linkages):

The increase in collaborative R&D activity may be relevant in broader human-resources science and engineering (S&E) issues such as training and career paths of S&E personnel. Linking firm-based databases on research joint ventures and university partners with, for example, university outcomes data on education and technology transfer activities, may provide a research tool for tracking the social impact of these activities.

The use of these data, however, implies the convergence of two different sets of disciplinary traditions and statistical cultures:

a) Industrial data (either establishment or company surveys): collected in terms of industry classification (NAICS, ISIC Rev. 3),
b) Education and occupational data (household or institutional surveys): data classified by field of study/field of science.[13]

5) **State/local** S&T data on industry-university collaboration and related local government activity (database linkages):

Explore the use of locally-developed data for analysis of national trends and exploit locality detail in national (NSF) data (e.g., R&D data by MSAs). Key issues in the former strategy include differences in survey definitions, coverage, and methodology across separate survey or data collection methods. Organizations such as the Association of University Technology Mangers (AUTM) and the State Science & Technology Institute (SSTI) may be instrumental in this effort.

6) **Post-alliance data** – S&T outputs and commercial activities (database linkages):

Post-alliance data are critical to track the impact of innovative activities of alliances. Data collection efforts in this area may be outside NSF/SRS mandate. However, existing economic databases could be exploited through linkages with available alliance databases. Databases of interest include:

- Financial and operating data – sales, value added, capital expenditures, etc. (U.S. Annual Census of Manufacturers)

- Mergers and acquisitions (M&As) and foreign direct investment (FDI) (U.S. Bureau of Economic Analysis, U.S. Federal Trade Commission)
- Bibliometric and patents data
- Subsequent public R&D funding
- Subsequent private venture capital funding activity

Chapter Eight

IMPLICATIONS OF THE DIVISION OF KNOWLEDGE FOR INNOVATION IN NETWORKS

Paul Quintas[1]
Open University Business School

1. INTRODUCTION

> Networks offer the benefit of both specialization and variety generation. [...] The converse of this statement is that firms are superior vehicles for the accumulation of specialized learning. (Kogut, 2000, p. 407)

In this chapter we approach innovation in networks and alliances from the perspective of knowledge processes. The central question is encapsulated in the above quotation from Bruce Kogut. Networks comprised of such individual firms each creating and accumulating specialized knowledge may, in an ideal world, benefit from the aggregate specialist knowledge of all the participating organizations. For each participating firm, their partners bring different knowledge from their own - i.e. they bring variety. There is a *division of knowledge* between participating organizations, implying lack of duplication. The central question is: how do they then understand, access or assimilate knowledge from their partners?

Drawing on current understanding of knowledge processes, a number of issues need to be taken on board if we are to address this question:

- Knowledge cannot be regarded as a commodity that is easy to trade and share, and the process of acquiring knowledge requires learning and understanding, which pre-supposes capability. 'Knowledge can no longer be pinned down to the heads of individuals and treated as a finished, stable product but is instead to be seen as a relational, transient product...' (Araujo, 1998, p. 324 building on Lave, 1993 and Lave and Wenger, 1991).

- Knowledge is created through dynamic processes. Knowing is an active process, resulting from action and engagement (Cook and Brown, 1999). Much knowledge is experiential and known to the individuals involved in specific activities. Experiential knowledge may be difficult if not impossible to communicate to others – much of it remains tacit.

- Knowledge is created in specific contexts and to an extent is dependent on context to acquire meaning (Lave, 1993; Lave and Wenger, 1991). Context specific (*or situated*) knowledge may be transient. Certain types of knowledge are 'sticky' and very difficult to share between contexts, even within a single organization. The concept of knowledge 'transfer' is problematic (in what sense is it given up by the originator?), and the knowledge 'flow' metaphor is particularly misleading.

- Knowledge has a social dimension. Communities of practice within which individuals share common work experiences and problem agendas provide a social context within which knowledge may be created and effectively shared (Brown and Duguid, 1998). Within the community of practice tacit knowledge may be shared in non-codified forms. (*ibid.*) Such communities have a degree of exclusivity – their knowledge is not readily available to outsiders who do not share similar practice-based experiences.

- The central paradox posed by the division of knowledge concerns the implications of specialization. One important way in which knowledge is advanced is through specialization (Loasby, 1999). In a network context, if each organization is pursuing specialized knowledge, how do the other organizations access this knowledge, or interface with it, without building similar capability? That is, if knowledge capability is effectively divided between the different partners, the capacity for knowledge sharing is minimal. If there is overlapping knowledge (or 'redundant' knowledge [Nonaka and Takeuchi, 1995]) between partners, advancement through specialization is reduced, and the benefits from network characteristics of difference, variety, and efficiency are diminished.

These are some of the considerations that we explore in this chapter, with particular focus on the latter paradox.

2. KNOWLEDGE, INNOVATION AND GROWTH

In spite of the importance placed on the role of knowledge in the economy by Smith, Marshall, Schumpeter, Hayek and others, Edith Penrose was to observe in 1959:

> Economists have, of course, always recognized the dominant role that increasing knowledge plays in economic processes but have, for the most part, found the whole subject of knowledge too slippery to handle. (Penrose, 1959, p. 77)

Nevertheless, major contributors explored the relationship between knowledge and firm growth (Penrose, 1959), the contribution of knowledge to the economy (Machlup, 1962), the nature of knowledge work (Drucker, 1969) and the notion of a knowledge-based post industrial society (Bell, 1973), but these issues went largely unnoticed by managers until the 1990s. From 1995 there has been extraordinary growth in interest in knowledge and how to manage it, across organizations in all sectors and in most industrialized countries. For the most part this interest has been driven by practical concerns.

Analysis of the drivers behind this interest suggests that a number of real-world factors have come together at this time (Quintas, 2002):

1) Wealth being demonstrably and increasingly generated from knowledge and intangible assets.
2) The rediscovery that people are the locus of much organizational knowledge.
3) Accelerating change in markets, competition, and technology, making continuous learning essential.
4) The recognition that innovation is key to competitiveness, and depends on knowledge creation and application.
5) The growing importance of cross-boundary knowledge transactions.
6) Technology limits and potentials: the limits of Information Systems and the potentials of communications and knowledge technologies.

These six factors came together in the early 1990s providing an impetus for practicing managers to focus attention on the challenges of managing knowledge and knowledge processes. All six factors are relevant to our theme, but two are central: innovation and cross-boundary knowledge transactions.

2.1. The Recognition that Innovation is Key to Competitiveness, and Depends on Knowledge Creation and Application

Increasingly in many sectors, not least due to global markets and increased international competition, competitive advantage occurs through innovation, whether in products, processes or services. In turn, innovation, which by definition is concerned with the new, depends on knowledge creation as well as application. Innovation is essentially a non-linear process, with management having to reconcile the predictable (incremental, cumulative) and the unpredictable (the injection of variety). Innovation requires eclectic and diverse approaches to the types and sources of knowledge that might be relevant. Innovation is about diversity, untypicality and uncertainty, and it therefore carries a considerable degree of risk. Innovating organizations must expect, recognize and learn from failure. In many sectors it is the capability to continuously create new knowledge rather than the possession of knowledge resources or assets that is the key to sustainable advantage, given the threat of innovation in the market place and the pace at which existing knowledge becomes outdated.

The management of innovation is therefore essentially about the management of knowledge capabilities as well as resources – the processes of creation, reformulation, sharing and bringing together of different types of knowledge. A central knowledge management dilemma for organizations seeking to innovate focuses on the need to build their knowledge bases cumulatively, share knowledge and learn from past experience, while at the same time generating variety and untypicality in their knowledge base as the source of novel developments. Drawing on Loasby (1999), Pavitt emphasizes that innovating organizations have to do two things:

> On the one hand they have to generate variety in order to foster the process of discovery of novel solutions by developing specialized bodies of knowledge. On the other, they have to co-ordinate dispersed learning processes that rely on different bodies of knowledge. (Pavitt, 2000, p. 13)

This co-ordination between distinct knowledge specialisms requires responsiveness, understanding and active management of relevant interfaces. (Pavitt, 2000) These are key knowledge management challenges.

2.2. The Growing Importance of Cross-Boundary Knowledge Transactions

Arguably no firm has ever been independent in knowledge terms, but it is certainly the case today that all organizations are increasingly dependent on external sources of knowledge. The complexity and pace of change in markets and technologies makes it impossible even for the largest organizations to cover all potential developments and to grow knowledge capabilities across the waterfront. This is exacerbated by the unpredictable nature of change in many sectors, from manufacturing to services.

A survey of 667 firms conducted in parallel by the Open University Business School as part of the ESRC *Innovation Programme* found that there has been a marked trend towards the outsourcing of R&D functions. Almost one third of the organizations surveyed have increased R&D outsourcing in the last five years, thus increasing the need for cross-boundary knowledge transactions when firms wish to innovate (Taylor *et al*, 2000).

Cross-boundary knowledge transactions apply to boundaries within organizations, between functional specialisms and between disciplines, but increasingly knowledge is accessed and shared across cultural and national boundaries as organizations and markets become international. In all but the most exceptional fields (nanotechnology perhaps being one) the vast majority of new knowledge is created outside (even the largest) corporate boundary. Ineluctably, therefore, organizations must develop absorptive capacity: the capability to access and assimilate new knowledge from external sources (Cohen and Levinthal, 1990). Knowledge interdependence creates new management challenges resulting from the risks and difficulties of knowledge transactions across boundaries. So too, the development of new products, systems and services increasingly requires the integration of knowledge from many disciplines. The ability to share knowledge across functional and disciplinary boundaries presents particular challenges since different communities and disciplines may have little common ground for shared understandings.

The awareness that innovation increasingly takes place across clusters or networks of organizations has been supported by a growing body of evidence (e.g. Tzokas and Saren, 1997; Taylor *et al*, 2000). Pavitt (2000, p. 11) emphasizes the 'need to cope with two fundamental attributes of innovative processes: (1) their embeddedness into a dense network of external relationships; (2) their combinatorial nature (in terms of the number of technical fields and components to integrate).'

Thus Loasby's (1999) privileging of specialization as the foundation of the advancement of knowledge is counterweighted by the resultant need for interactions between specialisms.

For some this new interest in knowledge management is an opportunity (some would say a requirement) to re-think the nature of organizations in knowledge terms, and indeed lay the foundations of a new theory of the firm (e.g. Spender, 1996; Grant, 1996). Our concern here is rather to identify the practical concerns, the key processes and challenges that face organizations and the managers who must make decisions.

As J-C Spender has observed there is no point in introducing the concept of 'knowledge' into management discourse unless we take the subject seriously and begin to utilize the perspectives it offers to provide new insights. For example, although the concepts of tacit and explicit knowledge are often used incorrectly as oppositional concepts (incorrect in the sense that Polanyi's (1966) formulation shows that all knowledge has a tacit dimension) nevertheless the raised awareness of the importance of tacit knowledge hopefully drives rather greater levels of critical understanding in relation to processes like knowledge sharing or 'transfer'.

2.3. The Nature of Knowledge

The advantage of introducing the concept of knowledge into discussions of networks and alliances in the innovation processes is that, providing we take on board the singular characteristics of knowledge, it leads us to ask a number of searching questions whose answers might otherwise be taken for granted. Of course, if we treat knowledge as an unproblematic commodity which can be packaged, traded and assimilated without difficulty then we will tend to miss the point. The disadvantage of introducing knowledge into our discourse is that the subject is complex and less than wholly amenable to simple solutions. In a short chapter we cannot hope to do more than hint at the epistemological concerns that are important and relevant. We therefore focus on the key issue at the root of much of the complexity for organizations - the tacit (or implicit) nature of much human knowledge.

This can be illustrated by analogy. Imagine if you will a group of jazz musicians who come together for the first time. All accomplished musicians, each a virtuoso on their own instrument, but specialists, not multi-instrumentalists. They know how to play their instrument because they have practiced, and internalized experiential knowledge. All are experienced improvisers who have performed with many different line-ups. They share a knowledge of the jazz canon, being steeped in listening to the greats over many years. They know how to play in a group and react to their fellow musicians' expressions, exchanging phrases and changing the tempo in an interwoven pattern that is created spontaneously. They have fun. They are, in effect, a community of practice, to which we will return a little later.

There are many learning opportunities in such a scenario, and indeed each musician may well take away with them new insights, phrasings, interpretations, and other learning points about group interaction. However, even with this level of empathy and shared reference points, the piano player cannot go home and play the saxophone like the virtuoso colleague in the group, and vice-versa. Neither could they learn each other's instrument by having extended conversations over the next six months about technique and method. No. In order to acquire knowledge of a new instrument, they would each have to practice. Such experiential knowledge cannot be codified – it remains tacit. In a sense what they possess is the ability to enact knowing, rather than knowledge – it is a set of cerebral and motor-neurone internalizations, which only have reality when they play. Heaven forbid, should anything happen to their hands, where then is their knowledge?

Similarly, organizations survive, innovate and prosper because people within them generate knowledge, and a proportion of this is experiential and must remain tacit. This awareness points to the limitations of attempts to manage knowledge based on that which is codified or codifiable. It is a simple point, and has profound implications, but it is one that is all too easily glossed over.

2.4. Knowledge and Innovation

Knowledge is an input to innovation, inseparable from the innovation process, and is also an output of that process. Innovation is concerned with the new or the novel, the break with the past, with change. Innovation and knowledge management are linked through four key processes: knowledge application, creation, transformation and integration. When we are innovating, we are also creating knowledge. Therefore the processes with which we are involved have two types of outcome: 1. the innovation itself; 2. the generation of new knowledge.

To say that new knowledge is created during the innovation process is to say that learning has taken place – those involved have collectively or individually learned how to do something new. Learning takes place throughout the innovation process, as problems are tackled, or knowledge and expertise is shared and integrated. This new knowledge may then be incorporated in the innovation, or may generate ideas for new projects and further innovation. These processes, however, are constrained.

According to Pavitt (1998) the knowledge base of large firms is cumulative and path-dependent, and constrains what they can learn:

> Large firms may have competencies in a number of fields of technology but, in the contemporary world of highly specialized knowledge, the costs of mastering all of them clearly appear to outweigh the benefits. Firms develop their technological

> competencies incrementally, and constrain their search activities close to what they already know. {...] The firm's knowledge base both determines what it makes, and the directions in which it searches (Patel and Pavitt, 1997). In this sense, there are clear cognitive limits on what firms can and cannot do. (Pavitt, 1998, p. 441)

This path dependency may be an appropriate incubator for continuous improvement and incremental innovation, but less so for step changes and more radical innovation. Where markets and products are relatively stable and trajectories of development predictable, these cognitive constraints may present less of a risk. Where, as is the case in many sectors and markets, change is less predictable, firms need to widen their knowledge base in order to be able to innovate. Cognitive constraints imposed by core capabilities may indeed become core rigidities (Leonard, 1995).

Cumulative and path dependent specialized development is essentially a linear response to the challenge of innovation. Whereas continuous improvement may conform to linear development characteristics, innovation with any pretensions to novelty has by definition non-linear characteristics. It requires the generation and or incorporation of variety in the development of knowledge (Kogut, 2000). However we should note that much of the knowledge generated through R&D is of a tacit nature and located in the specific context in which it was developed (Nelson and Winter, 1982). It may therefore be the case that knowledge is created but passes unnoticed.

Firms are faced with present and future decisions, without the benefit of hindsight available to post-hoc studies or innovation. Innovation is, by definition, characterized by uncertainty - since whatever confers the epithet 'innovation' (whether it is a radical change in technology, a new configuration of component elements, or a new context of application) it must be novel and not previously achieved. And while you are developing your next product drawing on your accumulated knowledge and path-dependent development, new rivals may be about to introduce a radical breakthrough that will outperform your product at lower cost. Decisions about which technologies to follow, which development pathways to pursue, run the risk of being cul-de-sacs that lead nowhere. Alliances, networks and collaborations provide the means by which firms can reduce the risk, share costs and scarce resources, especially with regard to currently 'peripheral' technology areas (Quintas and Guy, 1995).

2.5. Extra-Organizational Knowledge

While much influential work focuses on knowledge processes for innovation within the individual firm, there is also a sizeable body of evidence that extra-organizational knowledge plays a key role. In terms of knowledge resources and capabilities, few organizations would consider themselves to be a self-sufficient island. Such awareness underpins current emphasis on the importance of absorptive capacity or the organizations' ability to learn from external sources. As Cohen and Levinthal (1990) argue, absorptive capacity must be built.

Although discussion of knowledge transactions is a recent preoccupation for many students of innovation, the notion of technology transfer has a longer history. Indeed many authors writing of technology acquisition or transfer almost certainly assume implicitly that this includes the knowledge about the technology. Over a quarter of a century ago David Teece defined the cost of technology transfer as "the costs of transmitting and absorbing the relevant firm-, system-, and industry-specific knowledge" (Teece, 1976, p.36), thus defining technology transfer in terms of knowledge acquisition.

The question that arises is how far should firms develop, as part of their capacity to absorb knowledge, capabilities in the areas of knowledge they wish to acquire? When does absorptive capacity become an expanded internal knowledge base? Loasby follows his argument for the superior advancement of knowledge through specialization to its logical conclusion. He is clear that firms should not attempt to widen their capabilities to encompass emergent areas of knowledge:

> Many of the capabilities which a firm requires must be left outside its control if they are to continue to develop. Capabilities must remain distributed; what the firm needs is access to these capabilities. ... all firms, and all individuals, require absorptive capacity if they are to make good use of capabilities and information which is generated elsewhere. (Loasby, 1999, p. 96)

To do this, firms must be open to other ways of thinking; to be agile and receptive. The skills of absorption include techniques of sourcing, sense-making and acquisition. They must become adept at learning. However, much depends on the sector, the nature of the technology, the stage of the lifecycle of the technology and product markets, and the structural nature of relationships with external sources. We explore these issues in the following section.

3. NETWORKS, INNOVATION AND KNOWLEDGE

3.1. Networks

> Firms are not islands but are linked together in patterns of co-operation and affiliation. (Richardson, 1972, p. 895)

As Freeman (1991) shows, the external network has increased in importance as a source of resources available to the innovating firm. The key role of external linkages was noted a century before by Alfred Marshall, particularly, at that time, focusing on regional networks. As far back as 1951 Stigler linked networks with progress through specialization. Stigler (1951) observed that firms substitute inputs from external suppliers because these suppliers learn through specialization to provide these at lower cost than the vertically integrated firm can achieve.

Piore and Sabel (1984) argue that the network structure enables firms to fill specialist niches in an interdependent system of innovation and production. Miles and Snow (1994) propose the concept of the dynamic network of large and small firms interacting in collaboration. They place emphasis on the freedom to specialize this confers; for an individual organization, such a network relationship means that it can operate in its area of particular competence. This complementarity offers the opportunity to configure networks which can respond to complex situations and rapid adjustment to changing market conditions - referred to as flexible specialization by Piore and Sabel.

Networks vary, for example depending on the number of collaborating participants, and the configuration of their inter-relationships. These are influenced by each participating firm's access to resources, degree of specialization, location in the value chain, the extent to which information exchanges are standardized, duration, power relationships, and the level of inter-dependence in relations (Gelsing, 1992; Alter and Hage, 1993). Network structure also tends to vary across the industry life-cycle. So, for example, during innovative industrial phases, networks tend to be less formal and be populated more by geographically proximal members (apart from where there are strong relations between industry and research) than during more mature industrial phases (Gelsing, 1992).

For economists the network allows a configuration of resources that lies between markets and hierarchies. Between the extremes of market and hierarchy many kinds of formal and informal collaborative relationships between firms exist, from 'loosely coupled' to 'tightly coupled':

> ...what confronts us is a continuum passing from transactions, such as those on organized commodity markets, where the co-operative element is minimal, through intermediate areas in which there are linkages of traditional connection and goodwill, and finally to those complex and inter-locking clusters, groups and alliances which represent co-operation fully and formally developed. (Richardson, 1972, p. 887)

3.2. Transient and Persistent Networks

Freeman (1991) and others distinguish different types of network, especially informal versus formal. However one important discriminating feature appears to have escaped the attention of most writers, that of network duration, that is, whether any particular network is short-lived or persistent. Considering this dimension, two important conceptualizations of 'network' seem to be relevant to clusters of firms involved in innovation:

1) Transient networks: limited duration project or product based clusters of organizations brought together for a specific objective. Following completion each participating organization may reconfigure as part of a different network, constituted to meet different project objectives. This type of network is common in government and EU sponsored collaborative research projects, as well as straight commercial collaborations for specific problem-solving or product development.

2) Persistent (or durable) networks: a sustainable cluster of organizations which has a lifetime beyond a specific project. In many accounts the persistent network also has persistence as an entity even when organizations leave and new ones join (c.f. Rycroft and Kash, 1999). These networks are therefore dynamic and adaptive. In the view of some authors this suggests that the persistent network is a form of organization that has an identity independent of (and synergistically greater than) its constituent parts. This type is therefore associated with ideas around learning networks and self-organizing networks: e.g. "These networks are learning organizations ....." (Rycroft and Kash, 1999, p. 28). This approach considers the network to be an organizational entity which reconfigures over time, which learns rapidly, self-organizes, makes decisions e.g. 'Networks must be careful not to 'push the envelope' on too many frontiers of tacit knowledge at the same time' (Rycroft and Kash, 1999, p. 148) and indeed has many of the characteristics of an individual organization.

Of course any individual organization may be involved in both types of network either concurrently or sequentially. Also individual organizations may be 'serial networkers' participating in a series of transient networks. We should also acknowledge that the above distinction may be somewhat muddied by persistent networks of firms within larger transient networks: i.e. examples of closely collaborating firms that network together for more than one project and within more than one transient network.

The distinction between transient and persistent networks is of course important in relation to any discussion of network properties. For example in considering the notion that networks learn we need to reflect on the location of that new knowledge after a transient network has broken up. Conceptually, once we assume that the network is the unit of analysis, the organizational entity that makes decisions, has strategies, learns etc. independently from the organizations that make up its elements, it follows that we have to consider the possibility of conflict between the network and it's constituent organizations. What is best for the network (its learning, or its persistence) may not be what is best for an individual firm within the network, which may well have other businesses environments, and product markets and innovation priorities to address. The macro-economic policy implications of this are stark, and Schumpeterian. It matters not whether the individual enterprise survives as networks evolve. However, legislators and politicians may find networks more difficult to cope with than individual firms.

3.3. Innovation and Networks

There are of course many different types of innovation, and different types of network. The familiar distinction between incremental innovation, radical or discontinuous innovation, systemic innovation, and changes in techno-economic paradigm (Freeman, 1982) is ever relevant. Langlois (1992) usefully distinguishes between *modular* innovation, comprised of decomposable elements which firms may configure to provide new products, and systemic innovation which depends on the effective interaction between complementary capabilities. In broad terms, modular innovation may be achieved by market transactions between de-coupled firms (Pavitt, 2000) whereas systemic innovation requires a degree of collaboration and integration. Pavitt (2000) refers to the degree of integration as spanning a spectrum from *loosely-coupled* to *tightly-coupled.*

Modularity requires the innovation to be comprised of black-box modules, each developed by specialists. If interfaces are known and stable, then each partner organization only need know the input-output characteristics of the black box modules produced by others – they do not need to understand what is happening inside the black-box. Within each module, improvements can be made, up to a point, without requiring other modules to make changes. De-coupled arrangements are found where rates of change in component technology are low, and systems interdependencies (i.e. relationships between components) are predictable (Pavitt, 2000). 'A de-coupled system is a network without systems integrator' (Pavitt, 2000, p. 16). Langlois (1992) argues that informally organized market transactions best serve the achievement of modular innovation.

In more complex innovation contexts such as complex product systems, the decomposition of the whole into component modules simplifies tasks within each decomposed module, but, as Simon has shown, this shifts the locus of complexity to the interrelationships and interfaces between the elements of the system. Where component or module technology is changing significantly, or where system interdependencies are unpredictable (surely the case if the innovation has any degree of novelty) each organization in the network must retain in-house capability in order to understand and exploit the options emerging across the network (Pavitt, 2000).

The key role played by a systems integrator in many networks is illustrated by the example of Toyota. In Toyota's network, innovative capability is distributed. The network of suppliers evolved, 'to move the single locus of innovation from the core assembler (Toyota) to the suppliers as well' (Kogut, 2000, p. 421). This provides variety, 'based upon specialized competence' (Kogut, 2000, p. 422). Kogut refers to this as 'black box modularity' which 'permits specialization, and yet demands a high degree of coordination' provided by Toyota. (*ibid.*)

Already we can see different types of innovation correspond to different types of network, as is argued by Chesbrough and Teece (1996). They also consider autonomous innovations which may be developed within single integrated organizations. Chesbrough and Teece maintain that systemic innovations are best managed within a single integrated company, because dependence on external knowledge and information leaves the firm vulnerable especially if conflicts occur.

Single-firm innovation is not our main concern here, but it is important to note that the network form is not the only solution, nor necessarily the best. In a recent study into knowledge processes in the semiconductor industry, Almeida *et al.*(2001) concluded that the multinational corporation is superior to strategic alliances or market transactions in terms of knowledge sharing, creation and integration (referred to as 'knowledge building'). This study used patent citation data as an indication of knowledge building, supplemented by interviews to explore the processes involved.

Chesbrough and Kusunoki (2001) add a dynamic element to their analysis using the example of the hard disk drive industry. They suggest that disk drives went from single firm integration in the early stages when the technology was poorly understood and fast changing, to decomposed systems better provided by networks of module suppliers. This latter discussion serves to illustrate how different industries and different technologies may conform to different patterns, and indeed the appropriate organizational form, and locus of knowledge, is likely to vary over innovation lifecycles (Storper, 1996).

3.4. Network Linkages and Coordination

Pavitt (1998) notes that cross-boundary linkages between specialisms are poorly understood and cannot be thought of simply as 'flows' of codified information since tacit knowledge and ambiguity is unavoidable:

> In our present state of knowledge, effective coordination belongs in the field of practice rather than the field of understanding. [...] Experimentation and learning across critical organizational interfaces are particularly difficult when combining knowledge from different functions, professions and disciplines. (Pavitt, 1998, p. 444)
>
> Models of intra-corporate coordination have not got very far in grappling with these essential features of innovative activities. For example, in Aoki's models (1986), problems of coordination are... dealt with through information flows, rather than in learning and innovation mediated through tacit knowledge. (Pavitt, 1998, p. 444 footnote 19).

At the minimum, knowledge integration or sharing requires common knowledge (Grant, 1996) or redundant knowledge (Nonaka and Takeuchi, 1995). Grant (1996, p. 116) suggests that this common or redundant knowledge is comprised of different types that fulfill different roles in knowledge integration: a common natural language; symbolic communication using mathematics and computer code as well as literacy; overlapping specialist knowledge, shared meanings through tacit exchanges, the use of stories and metaphors, and recognition of other's knowledge domains through mutual adjustment.

Kogut (2000, p. 414) argues that the coordination between the nodes in a network may be via unique pathways or ties, or conversely via multiple ties. The first type draws on Burt (1992, 1997) who argues that unique or 'non-redundant' ties between nodes that are not otherwise connected may be provided by brokers who are able to gain advantage (or rents) because of their relative power derived from arbitrating the flows of information. Conversely, Coleman (1990) stresses the role of multiple or redundant ties as a solution to the problem of coordination between network nodes. In this case the benefits are derived from trust and shared learning resulting from long term relationships and good communication between nodes. Coleman networks are characterized by few nodes with dense ties (such as the Toyota example) whereas Burt networks have many nodes with few redundant ties, and many links furnished by brokers. The Coleman network structure parallels a network with a central 'systems integrator' with the capability to coordinate a range of differentiated network partners. According to Kogut (2000) dense (Burt) networks "provide an important capability of knowledge acquisition" (p. 421) by means of "rapid flow of competencies through frequent and on-going relationships" (p. 422).

4. A KNOWLEDGE PERSPECTIVE ON NETWORKS AND INNOVATION

4.1. The Division of Knowledge

The term 'division of knowledge' intentionally links the central issues discussed here to their antecedents in the 'division of labor'. "Adam Smith explained the wealth of a nation by the division of labor which led to the growth of differentiated knowledge." (Loasby, 1996, p. 299) The division of labor results in the division of knowledge, and specialization:

> ...knowledge grows by division; each of us can increase our knowledge only by accepting limits on what we can know. (Loasby 1999, p. 135)

Thus Hayek's identification of "a problem of the *Division of Knowledge* which is quite analogous to, and at least as important as, the problem of the division of labor" (Hayek, 1937, p. 49, italics added) is remarkable for his use of the word 'analogous' when the division of knowledge is central to Smith's focus on the division of labor as the basis of wealth creation through the development of specialized skills, expertise and knowledge: "Each individual becomes more expert in his own peculiar branch, more work is done upon the whole, and the quantity of science is considerably increased by it." (Smith, 1887, p. 12). Indeed Adam Smith presented empirical evidence that the division of labor leads through specialization to innovation: "...the invention of all those machines by which labor is so much facilitated and abridged, seems to have been originally owing to the division of labor." (Smith, 1887, p. 10)

The strong message that emerges from Smith and Loasby is that knowledge advances through specialization. Human minds cope best with focused and bounded scope, within which they can add depth of knowledge. Loasby emphasizes that the knowledge accessible to any individual is a minute part of the knowledge that exists globally, but even within bounded domains individuals' knowledge is limited: "Nobody knows how a Boeing 737 works; and nobody knows how the Boeing Company works" (Loasby, 2001, p. 7). This clearly raises important questions for communication and integration of knowledge across specialisms, at the level of the individual, the organization, and the network.

At the extreme, perfect specialization (no overlapping knowledge resources or capabilities) will maximize the costs of knowledge sharing. Increasing overlap (knowledge redundancy) carries costs of duplication and reduction of advancement through specialization. Grant describes the paradox, which is central to the organizational issues addressed in this paper, by analogy:

> ...if two people have identical knowledge there is no gain from integration – yet, if the individuals have entirely separate knowledge bases, then integration cannot occur beyond the most primitive level. (Grant, 1996, p. 116).

4.2. Context Specificity and Stickiness of Knowledge

Current perspectives on knowledge suggest that knowledge is created in specific contexts and is to varying degrees 'situated' (Lave, 1993) and may be 'sticky' (von Hippel, 1994) and difficult to transfer or share. It follows that in order to transfer knowledge between contexts it must be de-contextualized, or made independent of context.

Much of the knowledge generated by or associated with innovation is tacit knowledge that is held implicitly by those directly involved, and may also be embedded in the organizational and social processes that generated it. Explicit (codified) knowledge can be expressed and transferred in written and other recorded forms (reports, designs, formulae, specifications etc.) which may be understood by people with complementary knowledge who can extract meaning from the 'codes'. Even in the case of explicit knowledge there is a tacit dimension to the process of understanding or extracting meaning from the information.

Skilled members of a community of practitioners are often unaware of the tacit knowledge they possess – e.g. their problem solving behavior, the rules that they follow and the knowledge sources that they draw on. Moreover, communication among community members is often dependent on implicitly held knowledge of community-shared interpretations, beliefs and culture.

Tacit knowledge is personal, linked to experience and learning, and cannot be codified. In some situations tacit knowledge may however be shared (via non-codified pathways) within groups that share common learning experiences and understandings rooted in common practice. (Brown and Duguid, 1998)

Indeed, many authors argue that the only way to transfer tacit knowledge is through guided joint social interaction during, for example, apprenticeship within a community of expert practitioners. (Lave and Wenger,

1991) Similarly where knowledge associated with technology is being transferred between organizations: "in the absence of intimate human contact, technology transfer may sometimes be impossible." (Teece, 1981, p. 86)

In a revealing study of collaborative projects involving Western and Japanese firms., Hamel *et al.* (1989) found that Western companies tended to bring easily imitated technology to a collaboration, whereas Japanese firms' strengths were often 'difficult to unravel' competencies which were less transferable (Hamel, Doz and Prahalad, 1989).

4.3. Communities of Practice

A Community of Practice (CoP) is an informal social network in which the engagement of individuals in a joint enterprise is facilitated by a shared repertoire of concepts, actions, tools, stories, artifacts, and discourse (Wenger, 1998). This self-perpetuating group differs from other teams in that although there is a diversity of experience and background, all members, through extensive communication and shared practice over a period of time have come to share similar goals and interests, beliefs and value systems without any formal organization (Wenger and Snyder, 2000).

CoPs enable expertise to be shared and best practice to emerge, freely and informally. The commitment to the CoP overrides any reluctance to share knowledge, and the community, as it is cultivated, accepts the boundaries that their knowledge sharing activities fall within. It is this trust among the CoP that demonstrates its uniqueness and ability to go beyond the team and networking practices.

CoPs enable learning through shared experience, creating knowledge that is socially held. In the case of a newly developing CoP, all members of the community have different knowledge and expertise. Each member is differently knowledgeable about their own sphere of existing expertise and at the same time differently novice about others' areas, and about the context in which new practice-based expertise is emerging. Engagement in the new context of practice involves progressive negotiations between 'experts' and 'novices'. Each member of the community assumes the role of novice in order to learn about aspects of others' expertise and, simultaneously, to learn through engagement in practice.

Learning in an established CoP may also be a guided process. Guided joint social activity mediates experience, decontextualizing it in order to enable the sharing of knowledge. Thus another vital role of the CoP is to provide a support structure between those who have greater knowledge and experience in a particular subject than others. Such practice draws on individuals at all levels and enables reciprocity as new problems and issues emerge. It enables individuals to sound out new ideas/new concepts prior to putting them into practice in the workplace or other environment. Such sounding out is key to drawing on the experience and expertise of the members of the community, learning from other people's mistakes, what works and what doesn't. The CoP provides a safe haven.

While many CoPs exist within single organizations, they also may cut across organizational boundaries. This theme is explored below.

5. EVIDENCE FROM CASE STUDIES

We now present some findings from two research projects. The first project focused on R&D collaborations in the UK IT industry in the 1980s and unearthed a number of relevant issues. We then report on a series of case studies conducted in the late 1990s which looked at different types of innovating organizations involved in different forms of network relationships.

5.1. The Alvey Collaborative Pre-Competitive R&D Program

In the 1980s the UK government supported a major program of collaborative R&D into advanced information technology – the Alvey Programme. This program spanned over five years and involved the majority of organizations with R&D capability in information technologies in the UK. Research covering 198 pre-competitive R&D collaborations, involving 115 firms, 68 academic institutions and 27 government research laboratories (Quintas and Guy, 1989; 1995) found that firms have a range of strategic objectives when entering into R&D collaborations:

- To acquire know-how and enhance their knowledge bases, primarily to exploit this knowledge in the subsequent in-house development of technology, but also to inform strategic decision-making regarding technological capabilities;
- To supplement their skills and expertise bases through sharing scarce skilled personnel;
- To reduce R&D costs and share risks;
- To participate in collaborative development of standards, and to track and learn about emergent standards;
- To develop a national base of collaboration upon which to develop international links.

The study found that achievement of the first objective (acquisition of know-how and enhanced knowledge bases) depended on a number of factors:

- The capability and compatibility of collaborative partners in terms of expertise levels need to be roughly on a par at the start;
- The nature of the collaboration: i.e. how work is divided, mechanisms for integration, the channels of communication;
- The nature of the technology or knowledge being generated and transferred (e.g. codified knowledge such as software modules are easier to transfer than tacit knowledge);
- Stability within participant organizations in terms of ownership, business and technology strategy, and continuity of personnel active in the collaboration;
- The ability of the recipient firm to understand and internalize the technology and know-how, i.e. their absorptive capacity or ability to learn.

(Quintas and Guy, 1995, p. 341)

The incompatibility of knowledge and expertise was a key factor in many problematic collaborations. Participants would complain that they had lost significant amounts of time ‘bringing their collaborators up to speed’, or in some cases educating them up to a level where they could begin to communicate.

Conversely, there were many examples of successful collaborations where the project teams, comprised of individuals and small groups from many organizations, formed an effective inter-organizational community of practice. Members of these teams commented that their working relationships with project colleagues based in external organizations were far better than those within their own company. Colleagues in their own firm didn’t understand what they were working on, and they found they couldn’t communicate easily on a technical level with people outside the project team. Non-technical barriers grew between themselves and fellow employees (and some managers) who became suspicious and skeptical about a project that involved close collaboration with outside organizations.

This research led to the conclusion that at the strategic level a strong case for R&D collaboration can be made where the knowledge or technology is currently marginal to the firm’s core development trajectory. This may be because the knowledge or technology is new and its contribution to future product development uncertain. Collaboration enables firms to track such developing and risky areas without the costs of building strong in-house capability, at least until it is clear that this has become a business imperative. Conversely, even where collaborations are focused on R&D trajectories that are central to participating firms’ current business strategies, in the projects studied there were several costly examples of business strategies being changed and the reason for collaboration negated and routes to exploitation cut off.

This raises important questions about the ability of other partners in the network to exploit the outcomes of the collaboration. In those networks clustered around a systems integrator the satellite organizations are unlikely to have the capability to take things forward, other than spin-off outcomes in their own specialized areas. The research also highlighted major barriers to knowledge sharing between project teams and the rest of their organization.

At the level of the individuals involved in inter-firm collaboration, their success in forming a community of practice with teams from other organizations contrasted sharply with the barriers that grew between themselves and their colleagues in their own firms.

5.2. Knowledge Management in Design and Innovation Networks

In 1996 the UK Design Council funded the three year research project *Knowledge Management in Design and Innovation Networks* (KAMDIN). The project stemmed from the understanding that product design and innovation increasingly takes place across clusters or networks of organizations. The KAMDIN project featured a number of case studies, two of which are described below in some detail, three in brief form (names disguised).

5.2.1. Case Study 1: TeleMeter[2]

TeleMeter (pseudonym) was a project that aimed to provide remote, discreet meter reading services, linking individual homes with various utilities (water, gas, and electricity) using an existing telecommunications network, eliminating the need for human meter-readers. TeleMeter required the development of hardware and software connected to a telecommunications system and to the utilities. Case study research focused on knowledge sharing among the distributed community of competing and collaborating firms that made up the network. Telecommunications company Tel-Net (pseudonym) was a systems integrator within a network of software and hardware suppliers, and potential customers, i.e. gas, water and electricity utility companies.

The project required the management of a community of specialists from different knowledge domains, working for different companies, all contributing to the design of a complex system. The technical issues raised by different specialisms influenced the different possible solutions emerging. However, understanding the different technical interfaces is only the start of the management problem, as the final design represented the different political influences of this community of specialists in the context of existing, functioning systems (e.g. telecoms network, utilities' systems).

Significant management effort was spent on ensuring that Tel-Net's own technical expertise was not being duplicated amongst the project network of firms. That is, economic considerations dictated that there should *not* be overlapping or redundant knowledge in the network. Remarkably, although Tel-Net brought in a specialist software house to work on the project, with software engineers located at Tel-Net's own R&D department, no explicit attempt was made by Tel-Net to learn from these people. Opportunities to learn from the collaborators were lost because of this culture. Throughout the project there were no mechanisms by which the creation of knowledge within one specialist area would be of benefit to other specialists, so the communities remained compartmentalized, their outputs 'black-boxed'.

The Tel-Net culture placed a great emphasis on formal, explicit processes, ignoring the informal and non-explicit elements of the project. It was assumed that, for example, the delivery of a commissioned report transferred the knowledge created and progressed the project. This was not necessarily believed by those who wrote the reports.

Management of the project finally passed from Tel-Net's R&D department to a subsidiary systems integrator within the corporate group. Its core expertise is the creation and management of large software systems. This new systems integrator benefited significantly from the knowledge about the market created during the project. Following the TeleMeter project the subsidiary company sold a telemetry system in a different market - remotely read gaming machines, rather than utilities. This large contract was won because of experience and knowledge gained with the utilities during the Telemeter project. Again this was not accounted for by Tel-Net, nor was it (explicitly) predictable. The product that was eventually sold by the subsidiary company used telemetry and a network system in a different way from that originally envisaged.

The potentially high cost of learning across discipline boundaries emerged as a significant management and cultural issue. Communication was mostly contractual, formal and codified (including minuted deliverables). The only communications that were valued were formal and codified. It was assumed that knowledge would be made explicit and would be delivered according to formal agreements, so opportunities for peripheral knowledge creation and informal sharing of knowledge were lost.

The case paints a picture in which the innovation process, and the inevitable uncertainty associated with it, sits uncomfortably in a management context that assumes predictability and certainty of outcomes established as targets at the outset. Although there were several alternative routes, configurations and potential new products, there was no mechanism for reviewing the initial assumptions in the light of knowledge created during the innovation process. The linear management of complex innovation did not account for, and was not interested in, the potentially valuable by-products of knowledge created across the network.

5.2.2. Case study 2: SouthTech

SouthTech (pseudonym) is a long established manufacturer of environmental monitoring equipment which it supplies to defense and security agencies throughout the world. The company's core technologies, developed to detect and identify chemical warfare agents, are a complex integration of electronics and chemical processes, miniaturized within a portable package. This very high technology continues to be refined for the protection of defense forces, while also expanding into commercial, law enforcement and scientific applications in international markets. SouthTech is now a subsidiary of a larger defense, medical and aerospace company. It is an R&D centered company, with marketing taking a subordinate role. The case study focuses on the design, development and manufacture of portable chemical agent monitors, exported to the US and other countries. These developments became dependent on external sources of knowledge when a key individual left the company. The case is unusual in providing insights into military product developments and into the relationship between the company and the defense establishment, and detailed insight into the sources of knowledge and expertise in a highly technical and interdisciplinary science-based product development.

SouthTech routinely has to 'tell its suppliers what to do' - i.e. provide them with knowledge in order to enable them to meet its requirements. Materials supplied have to be further treated by SouthTech to achieve required performance. SouthTech mainly has to maintain levels of knowledge capabilities that encompass and even exceed those of its suppliers in the fields that SouthTech and its suppliers span. It is an example of an innovative company that has to retain and continuously develop all the required knowledge in-house. Exceptions to this have occurred however. In one key project the need to grow in-house knowledge about miniaturization was removed by substituting a modular product from elsewhere. This reduced the need to learn to the acquisition of knowledge about the interfaces, but not of the module.

New knowledge is introduced into the company by individuals, and through the literature, through licensed technology and through technological systems, like CAD. The firm is conscious that its human resource policy must encourage staff to be receptive and to accumulate the knowledge breadth that will add robustness to the company's knowledge base.

Attempts to capture and codify project knowledge are fundamental to US Department of Defense (DoD) contracts. All participants in a project are required to maintain detailed logbooks of all their activities. These are handed over as required deliverables. They are treated by contractors as a required hoop through which they must jump. However what is recorded meets the requirement for data without passing on anything of value: 'we don't ship understanding. Understanding is not for sale!'

The SouthTech case provides a further example of the limitations of codified knowledge, and a remarkable insight into a solution to the problem of systems integration in complex systems. A key individual, who we will refer to as Brown, left the firm during the product innovation process. He passed CAD drawings and 50 assembled prototype products exposing manufacturing issues to his successors. However the successor team couldn't progress the development because even with these starting points they didn't understand the lessons from the prototypes. These required interpretation in the light of the tacit knowledge accumulated in Brown's head. Moreover they couldn't integrate all the complex component technologies conceptually. This task was exceptionally difficult because the component technologies spanned nuclear physics, electronics and chemistry. SouthTech were forced to re-hire Brown as a consultant. They were well aware of his unique polymath capability, as expressed by a colleague:

> Brown's integration in one mind is also vortex-like. He pulls in knowledge from literature, from schoolboy recollection, from experiment. But he shuns teamwork for teamwork's sake. The team offers him nothing until he *needs* the aptitudes of an industrial designer to materialize his thoughts....

Brown was aware of his own ability to conceptualize the whole product, and the lack of that ability elsewhere: 'I had remained in total control, and had short-circuited all the knowledge flow problems of a conventional design and manufacture company. (SouthTech) does not have the advantage of integration within one mind'.

This emphasis on the role of a key individual as a singular systems integrator is not uncommon in the folklore of industrial legends. The millions of lines of computer code developed for the operating system for ICL's mainframe computers in the late 1970s / early 1980s was said to be held in the head of one individual, the only person who could conceptualize the whole system. However the SouthTech example is remarkable in emphasizing the ability of Brown to integrate the otherwise divided knowledge from three disparate fields.

5.2.3. Cameo Case Study 3: Cellbase

Cellbase (pseudonym) is a UK based world leader in the design and manufacture of specialist products and subsystems used in wireless (mobile) telecommunications systems. Their products are used exclusively in cellular base stations around the world. Cellbase has design and manufacturing facilities in Europe, North America and Australia, and they collaborate with client companies and systems integrators (e.g. Nokia, Ericsson, Motorola, Lucent, Cellnet, Northern Telecom) across the world. The fact that they are often the only non-US company in their field selected to work on advanced US military projects illustrates the level of their technology. They were one of the fastest growing UK companies of the 1990s; internal growth augmented by a series of major international acquisitions. They are of particular interest because they represent a company with unique knowledge and technology. Also they have exploited military technology in a commercial environment. The case raises questions about the value of collaboration, the protection of intellectual property, and the challenge of staying at the forefront.

5.2.4. Cameo Case 4: Artyfact

This case focuses on the toy design company Artyfact (pseudonym). Artyfact is managed by designers, but two changes in company ownership over the duration of the case study has resulted in major changes in working practices that have implications for knowledge management. Initially, Artyfact was a wholly owned subsidiary of a civil engineering company who took a hands-off approach to managing the firm. Artyfact existed as part of extensive collaborative networks that includes external specialist designers, toolmakers, and manufacturing companies in the UK and the Far East. This represented an established cross-boundary *community of practice* that worked well based on clear understanding of roles in the network. Artyfact's takeover by a US firm removed this community, and Artyfact was almost destroyed by the transition as they lacked the capacity to manage the newly enforced processes for the communication of knowledge between the US parent organization and the UK outpost.

5.2.5. Cameo Case 5: F1 Grand Prix Engineering

F1 Grand Prix Engineering (pseudonym) (GPE) manufacture and race Formula One racing cars. This is an extreme case of a complex system in which continuous technological innovation is an absolute requirement. Race by race, every car that competes is essentially a prototype – there are always differences from previous race. Innovation is primarily focused on incremental painstaking improvement based on huge amounts of data (especially feedback from testing and racing). Periodically F1 or their rival teams develop more radical or step-change innovations. The latter are often associated with changes in the regulations (i.e. the 'formula') or inputs from different knowledge domains, e.g. orthogonal knowledge from computing, aerospace or new materials developments, allied to creative ideas of key engineers. The quality and rapid availability of data and information, and rapid learning, are paramount. Knowledge sharing with tire and brake manufacturers is vital. These are the main suppliers apart from the engine. Both tires and brakes are, like the cars, in continuous development throughout a season. Their manufacturers depend on streams of accurate data from F1 GPE - there is a mutual dependency and relationship of trust since these partners are also working for rival teams. Like most Formula One teams F1 GPE do not produce their own engines. The relationship with the engine supplier is closer than customer-supplier, but collaboration is confined to close interaction on required engine output characteristics (primarily power and torque delivery), plus external packaging. How the engine manufacturer designs the engine to meet these requirements is not a matter for collaboration.

6. DISCUSSION AND CONCLUSION

The central findings of the above research raise questions about the network as a model organizational form. The significant problem that surfaces is that knowledge sharing between specialisms becomes difficult if not impossible, without specialization-bridging measures such as redundant or overlapping knowledge. This is present in F1 and its suppliers, in Cellbase, and in SouthTech. In the latter two cases these firms usually have to know more than their supplier networks, in order to achieve the inputs they require. In the TeleMeter case the benefits of specialization - the absence of knowledge duplication or overlap – were insisted upon by the systems integrator, and became a barrier to knowledge sharing.

Specialists have their own knowledge domains, language and culture. These are some of the defining characteristics of 'communities of practice'. For a community of specialists to learn outside of their specialisms is very expensive, indeed we find that the cultural divides between specialisms is greater than those between companies. This observation also points up a significant negative side to communities of practice, currently regarded as fundamental structures supporting knowledge creation and sharing. While communication and knowledge sharing *within* communities of practice may be highly effective, we have to consider the resulting constraints on knowledge sharing *between* communities of practice. Certainly the success of some Alvey projects in developing inter-organizational communities of practice was mirrored by their distancing from their own organization.

As Pavitt (2000, p. 12) points out, the structure and organization of a network "is not something firms *have*. It is something firms *do*." Thus sourcing knowledge from outside the boundaries of the organization requires considerable effort and capability. The findings suggest that it is not possible to make a clear distinction between the possession of absorptive capacity and the development of an in-house knowledge base. In order to track and understand external developments and then to assimilate knowledge or technology from outside the firm boundary, firms effectively need to grow knowledge – indeed this is historically why many firms undertake R&D (Allen, 1977; Cohen and Levinthal, 1989).

We would expect, in accordance with the findings of Granstrand *et al.*(1997), that large innovating firms not only maintain distinctive capabilities in core technology areas but also maintain a background capability in those technological fields in which firms in their network or supply chain are specialists. This enables them to coordinate (systems integrate), be knowledgeable evaluators and purchasers, and learn from these external sources. This was indeed one of the key motivating factors for firms in the Alvey program. The alternative, illustrated by the TeleMeter case, is to proceed under conditions of ignorance, not only of what is inside the 'black boxes', but also in ignorance of the knowledge being created during the innovation process.

The findings support the view that knowledge cannot be regarded as a commodity that is easy to 'transfer' between collaborating partners. Knowledge acquisition depends on the development of capability, although this may occur within the network community of practice without being internalized by the firms concerned.

There are two main possibilities in devising mechanisms to counter the *division of knowledge* problem. The first focuses on learning across specialisms, where organizations need to devote considerable time and effort into knowledge sharing across disciplines. We cannot expect our organizations to be populated by polymaths like Brown (in the SouthTech case), but we have to create cultures and frameworks that encourage and support cross-specialist learning. This requires specialists to alternate roles between novice and expert, in order to share knowledge. This process may well be inhibited by individuals' unfamiliarity with the novice role, and within organizations' management regimes by the ways in which people's achievements are measured. It is, however, what appears to happen fairly naturally in many collaborative projects. We observed in the Alvey case that project teams from many different organizations with different specialisms gelled as communities of practice engaged in joint learning.

The second (related) solution proposed by Nonaka and Takeuchi (1995) focuses on building redundant knowledge into teams. The principle is that in order for specialists to share knowledge they require overlapping knowledge. One way to achieve this is for each team to contain representatives versed in the others' specializations. This redundant (or duplicated) knowledge enables communication, knowledge sharing and learning, but undermines the economic case for flexible specialization, and reduces potential for advancement within specialized fields.

In both solutions there is a requirement for joint activity, for working together and gaining shared experiential knowledge. It remains the case that without such close working practice, knowledge sharing will rely on codified knowledge and much tacit learning will be lost. In the context of innovation processes, where knowledge creation is almost certain but unpredictable, to miss out on these learning opportunities is to risk missing the point. This is particularly the case where the original goals are not met, as in the TeleMeter case. Such a project might be thought a complete failure by those who are not aware of the knowledge created. The systems integrator was also disadvantaged by its managerial culture which relied on formal communications, and on contractual transactions rather than trust. Both these prevented awareness of knowledge gains and indeed new business opportunities.

A further key question emerges from our knowledge focus: what is the nature of the knowledge division in any particular innovation network. What would constitute entirely separate knowledge domains, and indeed how disparate can knowledge domains be, before communication problems become insurmountable? Within the cases discussed above, it is notable that within Alvey collaborations, all focused on information technology, partners in collaborations often came from quite distinct knowledge areas, about which their partners knew little. Participants often referred to different vocabularies, and different ways of working, as well as clear examples of divided knowledge. This was more obvious in the TeleMeter case, and indeed was designed into the selection of partners, aimed at preventing duplication. Here the 'black-box' approach only worked up to a limited point. In the SouthTech case, the innovation brought together quite disparate knowledge domains (electronic hardware and software, nuclear technology, and chemistry). In this case only one person was able to bridge across this diversity, as was clearly shown by what happened when he left. For the rest of SouthTech the knowledge domain gap was clearly insurmountable. For F1, the organization has no choice but to build internal capability to bridge the knowledge gaps between engineering, aerodynamics, electronics, new materials, tire technology, etc.

The above discussion is based on a very narrow set of case studies, and generalisability into wider contexts cannot be assumed. Further work is needed to explore the organizational and management processes that emerge as being crucial. High level (firm strategy and network level) studies are useful, but limited. There is a dearth of studies of knowledge processes at the level of the innovation practitioners. Only by engagement with practitioners do we begin to understand the nature of knowledge creation, sharing and application in innovation contexts. In particular we need to understand the condition under which knowledge specialization will lead to severe communication problems, identify the costs of bridging these gaps, and explore the alternative solutions in a wide range of contexts.

A major methodological challenge remains in the area of tacit knowledge. Because of the nature of such knowledge it is not available to researchers (in much the same way it is not available to collaborating partners). Avenues to explore include ethnographic studies relying on long term observation and participation, and behaviorist approaches which seek to identify the symptoms of knowledge that betray its existence (if a jazz saxophonist can play, then we can assume he or she has internalized the knowledge).

There are also implications for policy makers. The network form of organization, collaborations and alliances, have all been encouraged by policies, and indeed by funded programs, in Japan, North America and Europe. The issues raised here around the division of knowledge should at least stimulate debate as to the assumed efficiency and effectiveness of networks as a solution in all areas.

PART IV

IMPACTS

Chapter Nine
ORGANIZATIONAL REQUIREMENTS FOR THE INNOVATION OF COMPLEX TECHNOLOGIES

Don E. Kash, Robin N. Auger and Ning Li[1]
Center for Science and Technology Policy
School of Public Policy
George Mason University

1. INTRODUCTION

Networks, alliances, and partnerships are essential to the innovation of complex technologies.[2] As used here, networks refer to organizational systems that exist to carry out the innovation of technologies. Networks are composed of multiple organizations that, at a minimum, range from the holders of core capabilities, to the suppliers of complimentary assets, to user organizations (see Figure 1). Alliances and partnerships are organizational relationships (of varying structure) commonly used to establish critical linkages in the networks that carry out the innovation of complex technologies.

Networks involve linkages and relationships that make it possible to access, create, synthesize, and diffuse the diverse tacit and codified knowledge necessary for the innovation of complex technologies. No organization acting alone has the capability to carry out the innovation of complex technologies. Networks, alliances, and partnerships are essential to contemporary technological/economic success because complex technologies are the most valuable technologies in the global market.

There is a positive sum relationship between complexity and technological innovation. Increasing complexity and increasing opportunities for innovation are inextricably linked; they co-evolve. Today's technology market is significantly structured by innovation-to-obsolescence cycles. Innovations create and expand markets, and they reduce and sometimes kill markets for existing technologies. Thus, the key to economic success for both companies and countries is the capacity to carry out repeated innovations ahead of or in parallel with their competitors. Only networks provide this innovation capacity.

Figure 1. Organizational Network

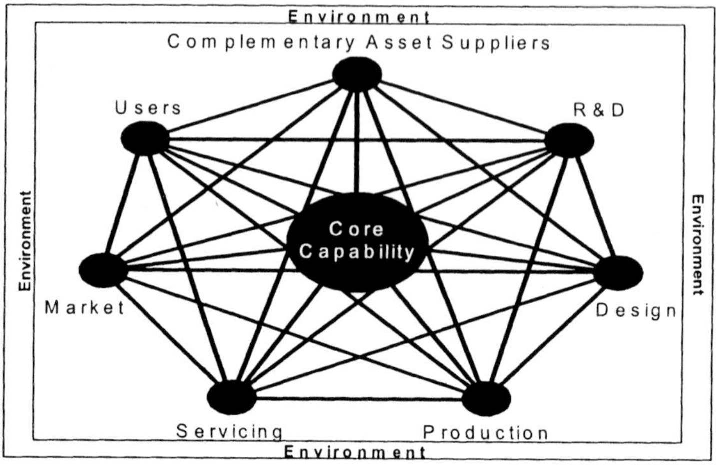

This chapter investigates: 1) three patterns of technological innovation, 2) the role of organizational networks in accessing and using knowledge, and 3) the decision-making associated with the different innovation patterns. Much of the chapter's source material is derived from five case studies of evolving technologies carried out in the United States, Japan, Germany, India and China. The concluding section suggests lessons for managers and public policy makers concerned with facilitating the innovation of technologies.

2. THE IMPORTANCE OF COMPLEX TECHNOLOGIES

The networks that carry out the innovation of complex technologies warrant major attention because the products and processes they produce represent 85% of the value of the world's thirty most valuable manufactured exports. Simple technologies—mostly shoes, clothing and toys—comprise the remaining fifteen percent (see Table 1). The thirty most valuable technology exports represent 44% of the value of all goods exports. There is a doubtless apocryphal story about a very successful bank robber who, when caught, was asked why he robbed banks. He said: "Because that's where the money is." The capacity to repeatedly carry out the rapid innovation of complex technologies is "where the money is" today.

Table 1: Thirty Most Valuable Manufactured Goods Exports, 1997[3] ($ Millions)

SITC	Commodities	Exports	% of World Total Exports
	Simple		
667	Pearl, prec-, semi-p. stone	42438.4	0.80%
684	Aluminium	43932.3	0.82%
699	Base Metal Mfrs. NES	40565	0.76%
821	Furniture, Parts thereof	50907.5	0.95%
843	Women's Outerwear Nonknit	43765.2	0.82%
845	Outerwear Knit Nonelastic	43599.1	0.82%
851	Footwear	43612.6	0.82%
894	Toys, Sporting Goods etc.	44079.7	0.83%
Subtotal		**352899.8**	**6.61%**
	Complex		
541	Medicinal etc. Products	83627.7	1.57%
583	Polymerisation etc. Products	74484.2	1.40%
598	Miscel. Chemical Products NES	39186.7	0.73%
641	Paper and Paperboard	70475.8	1.32%
674	Irn, stl univ., plate, sheet	51066.9	0.96%
713	Internal Combustion Piston Engine	58966.2	1.10%
714	Office Machines	36471.4	0.68%
728	Other Machinery for Special Industries	62815.5	1.18%
741	Heating, Cooling Equpt	39059	0.73%
749	Nonelectric Machine Parts, Acces NES	59913.7	1.12%
752	Automatic Data Processing Equpt	159009.2	2.98%
759	Office, Adp. Machine Prts., Acces	106831.6	2.00%
764	Telecom Equpt, Pts, Acces NES	141353.3	2.65%
772	Switchgear etc. Parts NES	70489	1.32%
776	Transistors, Valves etc.	198983.3	3.73%
778	Electrical Machinery NES	79732.2	1.49%
781	Pass. Motor Vehicles exc. Buses	259426.8	4.86%
782	Lorries, Spcl Mtr Vehicles NES	55670.4	1.04%
784	Motor Vehicle Prts, Acces NES	128014.1	2.40%
792	Aircraft etc.	92596.8	1.73%
874	Measuring, Controlling Equpt	60799.7	1.14%
893	Articles of Plastic NES	53978.3	1.01%
Subtotal		**1982951.8**	**37.15%**
Total		**2335851.6**	**43.77%**

Source: Compiled from United Nations, 1999 International Trade Statistics Yearbook, Volume I: Trade by Country and Volume II: Trade by Commodity, New York: United Nations, 2000.

3. NETWORKS AND SYNTHESIS

Successful organizational networks are distinguished by the fact that they can carry out innovations absent of full understanding of how they do it. Understanding requires the capacity to identify a system's key variables and the cause and effect relationships among them. When understanding exists, innovators know what actions will deliver the desired results. In science, theories provide understanding. Henry Ford understood his assembly line. However, no one can understand how complex technologies are innovated. For example, the innovation of the microprocessor utilizes too many sources of knowledge and has too many interactions to be understood completely.

Synthesis is a network core capability essential to the innovation of all complex technologies.[4] Synthesis is not understood; it is a capability networks develop, in part, through experience, i.e., trial and error. The most successful networks are those that can learn and adapt in the process of carrying out synthesis ahead of their competitors.[5] Network structures and processes that facilitate learning and adaptation are as important as capital, R&D, and intellectual property protection to the successful innovation of complex technologies.

In the five case studies summarized in this chapter we have sought to trace the development of networks, and their capabilities of synthesis, over the history of the five technologies. We have also sought insight into the relationship between national and company cultures in facilitating or impeding the development and adaptation of networks as they carry out the synthesis essential to innovation.

4. PATTERNS OF INNOVATION

Although it is not possible to understand the cause and effect relationships essential to the innovation of complex technologies, it is possible to associate technologies and innovation processes with one of three patterns (see Figure 2). The patterns are distinguished by different kinds of innovations and by the ways in which the networks access, create, and utilize knowledge and make decisions in carrying out the innovations. The first pattern—transformational—is distinguished by the innovation of first-of-a-kind technologies. The transformational innovation pattern is represented in Figure 2 by the square at the beginning of the trajectories. When the first-of-a-kind technology is complex, a network of organizations that range from complementary asset suppliers to end-users must be formed in order to incorporate the diverse knowledge needed for the innovation. Thus, within the transformational innovation pattern both a new network and a new technology must be designed and built, in part, through a trial and error learning process.

Figure 2. Technological Innovation Patterns

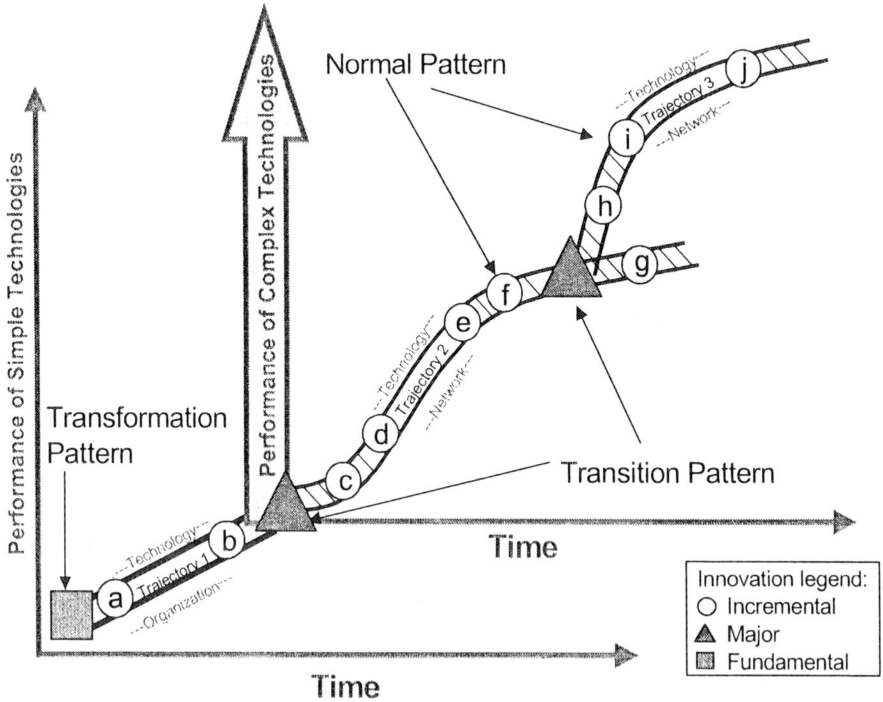

The second pattern—*normal*—is distinguished by innovations that deliver enhancements of established technology designs, for example, the ever more powerful microprocessors emerging from incremental innovations. The normal innovation pattern is represented in Figure 2 by the circles on the trajectories. Incremental innovations are predominately the product of cumulative learning within the network that was established to innovate the first-of-a-kind technology. Normal innovations also require diverse knowledge and participation by organizations that range from complementary asset suppliers to end-users. However, during the normal innovation pattern the linkages among the participants in the network already exist, and the network is able to generate internally most of the knowledge needed to carry out the incremental innovations.

The third pattern—*transitional*—is distinguished by innovations that deliver major redesigns, for example, the synthesis of electronics and mechanics. The transitional innovation is represented in Figure 2 by the triangles that connect the trajectories. Like the technologies, the networks that carry out transitional innovations must undergo major redesigns. The most distinctive characteristic of the network redesign is that new core capabilities must be integrated or merged to create a new network with the needed innovation capabilities. Thus the redesigned network must be able to access, create, and synthesize knowledge that did not previously exist in the network.[6]

Figures 3 and 4 show the development trajectories and different innovation patterns of five technologies produced by organizational systems centered in five different countries. Figure 3 traces the development associated with the Bosch (Germany) diesel fuel injection technologies.[7] Figure 4 traces the developments associated with the Sony (Japan) audio compact disc (CD),[8] the Hewlett-Packard (HP) (United States) cardio-imaging technology,[9] the TATA Consultancy Services (TCS) (India) software services and products,[10] and the Haier (China) line of home appliances.[11]

Figure 3. Bosch Innovation Trajectories

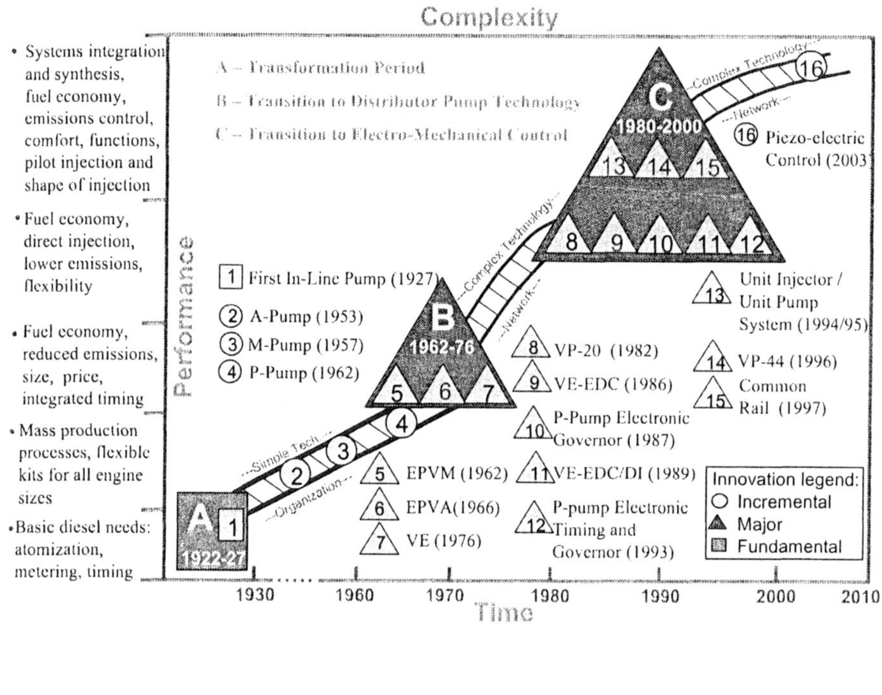

Some complex technologies have evolved from simple technologies, as is the case of the Bosch diesel fuel injection technology (Figure 3) and the TCS software (Figure 4). Some first-of-a-kind technologies arrive as complex systems, as is the case of the Sony CD and the HP cardio-imaging technology (Figure 4). Some technologies have not crossed the threshold from a simple to a complex technology, as is the case of the Haier home appliances (Figure 4).

Figure 4. Innovation Trajectories of Sony CD Systems, HP Cardio-imaging, TCS Software, and Haier Appliances

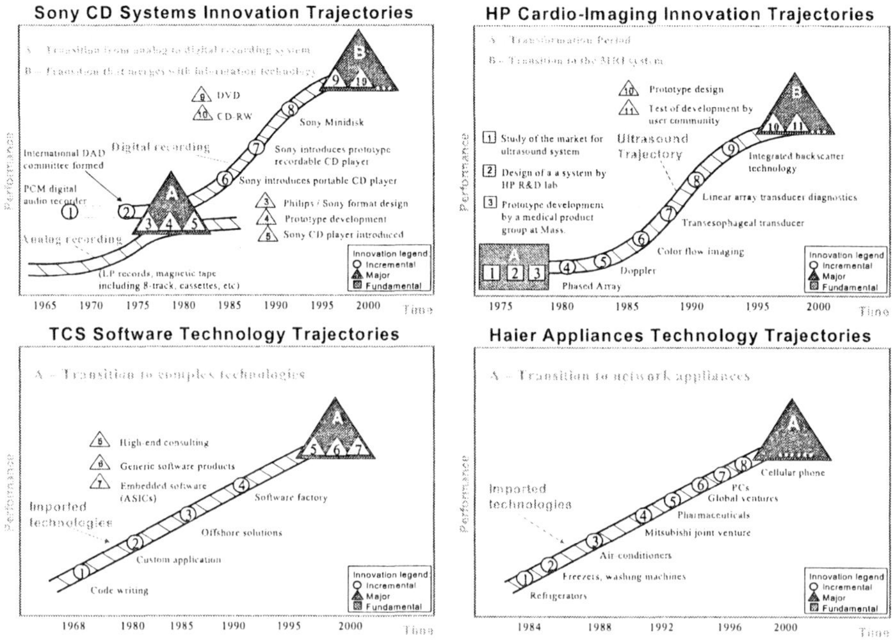

Simple technologies do not require organizational networks for their innovation; thus, the innovations of Haier have involved adapting technologies acquired from other companies to fit the Chinese market. Complex technologies always require organizational networks. One of the most difficult challenges faced by companies and countries occurs when it is necessary to make the organizational adaptation from the innovation of simple to the innovation of complex technologies. The simple to complex adaptation requires replacing a hierarchical organization that manages a narrow range of knowledge with a network that synthesizes a diverse range of knowledge. Bosch carried out this adaptation, for example, when it evolved from the innovation of all mechanical technologies to electro-mechanical technologies. This was a period in which a network had to be built that could synthesize electrical, electronic and mechanical knowledge.

For those networks already successfully carrying out the innovation of complex technologies, the major adaptation challenges occur when it is necessary to move from one innovation pattern to another. These adaptations commonly require either the integration of new organizations with new core capabilities into an existing network or they require the merger of existing networks for the purpose of producing a major technology redesign. This is evident in all of the case studies except that of Haier.

In the following section we summarize the diverse knowledge accessed and synthesized by the networks within each of the innovation patterns experienced by the five technologies we have studied. We also characterize the nature of the decision-making associated with each pattern within each study.

4.1. Transformational Pattern

Two of our studies include transformational innovations: the Bosch diesel fuel injection technology case and the HP cardio-imaging technology case. In the Bosch case the transformational innovation pattern encompassed the period in which a simple mechanical technology was produced. In the HP case the innovation pattern encompassed the period in which a complex electro-optical-mechanical technology was produced.

The Bosch innovation resulted from the identification of a market opportunity through personal interactions between Robert Bosch and the designers of a new engine—the diesel. An initial injection technology design was purchased from a Swiss company and skilled mechanical craftsmen within the Bosch workshop produced a simple mechanical design for a diesel fuel injection system that remained the dominant design for nearly four decades.

The HP cardio-imaging innovation resulted from a search for a new market for the company's medical electronics core capabilities. A market consultant hired by HP recommended development of the cardio-imaging technology even though most of the potential users—medical doctors—envisioned little need for it. The first-of-a-kind design required developing a number of first-of-a-kind subsystems (e.g., phased array transducer). A network was created that linked the HP R&D laboratory to the manufacturing facility and those two organizations to "lead user" medical doctors and subsystem suppliers. Knowledge that ranged from microwave radar to digital electronics and from signal processing to cardiovascular diagnosis had to be synthesized. Once a workable technology had been designed, it was necessary for the network to link with medical training facilities, medical schools and hospitals in order to train the doctors and technicians who would be the end-users of the technology.

The uncertainty faced by the HP organizational network included the specification of appropriate performance criteria and a lack of understanding of how to carry the design into development, manufacturing and market use. While the decision to innovate the cardio-imaging technology and the creation of the first network was undertaken by HP management, the technology development itself was undertaken by a network of people and organizations that had to synthesize diverse tacit and codified knowledge. Thus, a consensus process characterized the decision-making associated with developing the design. Development of the first-of-a kind design occurred more slowly than would have been the case had HP linked with an organization that had in-depth radar knowledge. HP had to acquire a body of tacit radar knowledge through trial and error that could have been obtained rapidly by establishing a partnership with an organization that held the needed expertise.

4.2. Normal Pattern

The normal innovation pattern, where established networks carry out incremental innovations that enhance existing designs, is the most characteristic pattern evident in all five case studies. Innovations that occur within the normal pattern produce the major economic payoffs for complex technologies, as innovation-to-obsolescence cycles structure market success for complex technologies.[12] Each incremental improvement creates new demand and reduces or kills the demand for existing technologies. The key to success, then, is to carry out the incremental innovations ahead of or in parallel with competitor networks.

Within the normal pattern the lack of understanding that networks must deal with concerns how to accomplish the next innovation. Alternatively, there is seldom lack of understanding concerning what the performance characteristics of the next innovation will be. Perhaps the most widely recognized example of this is "Moore's Law" in the semiconductor industry.[13] The networks that carry out normal pattern innovations have high levels of confidence that they will be able to accomplish the next innovation objectives. This confidence is brought about from the networks' repeated experiences in carrying out incremental innovations. When members of the networks are asked why they have such a high level of confidence in their ultimate success, they say, "Because we are working on it." The networks have confidence because they have developed network routines and heuristics that guide their problem solving trial and error activities.[14]

All of the case studies include multiple incremental innovations. In the Haier case, the incremental innovations involved a simple technology, the associated decision-making was centralized (restricted to company managers), and the knowledge needed to carry out the innovations was narrow. In two of the cases, Bosch and TCS, some of the innovations were concerned with simple technologies and some with complex technologies. Like the Haier case, the simple technology innovations involved centralized decision-making and the use of a narrow knowledge base. For the incremental innovations of complex technologies, the decision-making required broad participation of those holding the diverse expertise needed to carry out the synthesis required in producing the innovation. This same decision-making and knowledge pattern applied to all of the incremental innovations associated with the HP and Sony cases.

4.3. Transitional Pattern

Four of our case studies include transitional innovations: Bosch, Sony, HP, and TCS. In the Bosch case there were two transitions: one from a mechanical in-line pump to a mechanical distributor pump and one from a mechanical injection system to an electro-mechanical system. Between these two transitional innovations the technology crossed a threshold from simple to complex and the organizational structure evolved from a single hierarchical organization to an organizational network.

In the Sony case, the transition was from a complex electro-mechanical system (i.e., analog magnetic tape audio system) to an even more complex electro-optical-mechanical system (i.e., digital audio CD). The transition was the result of a consensus in the audio community that a major redesign was necessary. Two developments brought about the consensus: a saturation of the market for tape systems and the rapid development of digital electronics. Faced with these developments, the industry, encouraged by the Japanese Ministry for International Trade and Industry, evolved a consensus that the choice of the redesign should be made by an industry-created committee composed of the major manufacturers and not by a market competition. In need of a transitional innovation, Sony entered into a partnership with Philips (Netherlands) to design an audio CD. This innovation required the merging of two networks, where Philips had the leadership in optics and Sony in audio systems. The decision to establish the partnership was made by the companies' top management. The design was accomplished by a joint Sony-Philips team that made decisions by consensus.

The industry committee selected the Sony-Philips design (from three competing designs) as the industry standard. At that point Sony and Philips terminated their partnership and each created its own network to carry the design to the market. The Sony network linked with suppliers of subsystems, such as lasers, as well as to recording industry companies. The network utilized consensus decision-making processes to carry the design to the market.

In the HP case, the transition was from a complex ultrasound-based system to an even more complex magnetic resonance imaging (MRI)-based system. The transition resulted from market saturation of the ultrasound system and the capacity of the MRI system to provide diagnostic data that previously could only be obtained from physically invasive techniques, e.g. inserting a fiber optic tube in the vascular system. This capacity would create a new market for the major redesign of the technology.

The decision to pursue the MRI-based redesign began within the HP ultrasound network. A search was undertaken to find a partner with MRI capabilities. In this sense MRI capabilities required an already existing network that could put together the complex subsystems necessary to make the overall system work. HP recognized from the beginning that making a MRI system work required substantial tacit knowledge. While there was confidence that the existing HP network could gain the tacit knowledge through trial and error, it would take too much time in an environment where competitor networks were working on the same design concept. Consequently, the management of HP's medical group signed an agreement that merged its ultrasound network with Philips (Netherlands) established MRI network. The design of the new system was carried out by a team composed of people from the two networks. This merger differed from the one associated with the CD (discussed above) in that it called for the merged network to continue throughout the period of incremental innovations that would occur over the life of the redesign.

In the TCS case, the transition has involved moving from a simple technology and single organization primarily concerned with applying software expertise to the problems of specific customers to the development of networks that exist for varying periods of time that merge substantive industry knowledge with information technology expertise in order to provide high level consulting services, as well as to design embedded software—software used on application specific integrated circuits (ASICs). In its early period TCS supplied people who worked as subcontractors to write software code as one component of larger scale problem solving teams. TCS is now assembling teams that link computer makers and software product providers, such as Oracle, with industry-specific expertise to provide companies with high level consulting. Decision-making has evolved from being centralized to utilizing solution-specific teams assembled to meet specific client needs.

4.4. An Example: The Bosch Trajectories

The most complete illustration of the innovation patterns offered by our case studies is provided by the Bosch diesel fuel injection technology. The three patterns of innovation, plus the crossing of a threshold from a simple to a complex technology, are all evident in the Bosch case. Bosch began work on a diesel fuel injection system in 1922. Figure 3 provides an overview of the evolution of technological innovations associated with Bosch injection systems from the beginning to the present.

It took Bosch five years to develop its first commercial product, thus the transformational innovation pattern characterized the period from 1922 to 1927 (as represented by the square A in Figure 3) and was devoted to developing a dominant design, an in-line plunger pump, that was successful in the marketplace.

The normal innovation pattern characterized the process that produced enhancements in the in-line mechanical design after 1927. This pattern is indicated by numbers 2, 3 and 4 in Figure 3. The incremental innovations that produced these enhancements to the dominant design were primarily concerned with adaptations needed for engines of different sizes and with different performance requirements. What resulted was the creation of "kits" of components and design modifications that gave the in-line design significant flexibility.

The transitional innovation pattern represented by triangle B involved a major redesign of the injection system. The redesign produced an axial-piston distributor pump (number 5 in Figure 3). An alternative distributor pump design that used a hydraulic governor (number 6 in Figure 3) was also introduced during this period and had limited commercial success. A modified axial design, the VE pump, became a major commercial success (number 7 in Figure 3).

A normal innovation pattern, represented by the trajectory between triangles B and C, came next. It involved incremental innovations of both the axial piston design (VE) and the in-line mechanical pump (P-pump), plus innovations in process technology.

The roughly two decades between the two transitional innovations represented by triangles B and C is a period in which the integration of a diverse set of activities produced a complex technology. This period witnessed a major evolution from a hierarchal organization that utilized primarily mechanical knowledge to a network organization that utilized knowledge from many areas of technical specialization. For example, major electronics expertise had been integrated into the Bosch network in the mid-1980s.

The transitional innovation pattern represented by triangle C occurred in two phases. The first involved major redesigns of many of the mechanical injection systems. Solenoid actuator controlled electronic timing and electronic governors were integrated into the previously mechanically controlled distributor and in-line pumps, resulting in the production of five complex electro-mechanical injection systems in the first phase of the transition (see numbers 8-12 in Figure 3). In the second phase, three new designs provided greater capabilities for synthesizing diverse data and offered more precise control and higher injection pressures—the unit injector system/unit pump system (UIS/UPS), VP-44 radial piston distributor pump, and the Common Rail System (numbers 13, 14 and 15 in Figure 3)—carried the transition to electro-mechanical injection systems to its conclusion During this transitional period, a network had been developed that had greatly advanced electronic, electrical, software, and mechanical knowledge and had learned how to synthesize this diverse knowledge.

By 2000 the network that had emerged to innovate the electro-mechanical injection system manifested a normal innovation pattern—it repeatedly and rapidly carried out incremental innovations. The network was able to routinely innovate designs that could be orchestrated to meet the particular niche needs of diesel engines ranging from those used on lawn mowers to ship engines.

5. KNOWLEDGE AND DECISION-MAKING

Networks are essential to the innovation of complex technologies because they provide the capacity to do what neither conceptual systems (theories) nor individuals can do. Networks offer a way to access and use diverse knowledge that is both codified and tacit in order to accomplish what has not been done before without an understanding of how to do it.[15] Networks thus offer two critical capabilities. First, they provide a way to access, create, and use (synthesize) diverse knowledge that is located in multiple organizations and individuals. Second, they provide the structure needed for the range of people and organizations that hold the diverse knowledge to participate in decision-making.[16]

The Bosch case study provides an illustration of how knowledge/decision-making processes occur. In addition, the case study offers insight into both the difficult adaptation associated with crossing the threshold from a simple to a complex technology and insight into the structure and processes associated with each of the three innovation patterns.

A review of the knowledge needed to carry out the innovations of Bosch's diesel fuel injection technology over the last eighty years indicates a trend toward knowledge diversity that has been accelerating at an ever more rapid rate since the 1960s. The growing diversity of knowledge has two dimensions: it includes both greater knowledge depth (e.g., a capacity for executing ever finer machining) and breadth (e.g., the inclusion of electronics into a mechanical technology). The Bosch organization's reservoir of both codified and tacit knowledge has increased over its lifetime, and that increase has been accelerating in the last two decades. The knowledge is located in a growing number of organizations, both in other Bosch units and in organizations linked to Bosch through the diesel fuel injection network. The organizational network has developed a body of routines and heuristics that are essential to the successful interactions of the different organizations. The routines and heuristics exist as both codified and tacit knowledge embedded in the Bosch-centered network.

A review of the decision-making associated with the injection system innovations beginning in the 1960s suggests what is increasingly a self-organizing system—a system where decision-making occurs via a process that is coordinated by consensus concerning what the performance characteristics of the new innovation will be. Further, this self-organizing system is delivering decisions ever more rapidly. This was especially evident as the network entered the new century and was manifesting the capacity to deliver specialized configurations of multiple injection system designs. The exceptions to this general pattern occurred when the decision was made to design the mechanical distributor pump in the early 1960s and again with the decision to design an electro-mechanical injection system in the early 1980s. In these instances the top management of the Bosch company mandated both technical and organizational changes.

A picture of the knowledge and decision-making adaptation over the last eight decades is shown in Figure 5. Box 1 in the upper left corner of Figure 5 covers the period from 1922-1960 and reflects the fact that the knowledge needed to carry out the innovation of the in-line pumps was predominately mechanical and the decision-making was centralized in a few organizational leaders. Both the knowledge needs and the centralized decision-making during the first four decades reflected the fact that the technology was simple. The same knowledge and decision-making characteristics were evident in both the first-of-a-kind innovation taken to the market in 1927 and the follow-on incremental innovations (see Figure 3). In sum, for this simple technology the same knowledge and decision-making characteristics appear in the transformational innovation pattern and the normal innovation pattern.

Figure 5. Evolution of Decision-making & Diversity of Knowledge In Bosch Network

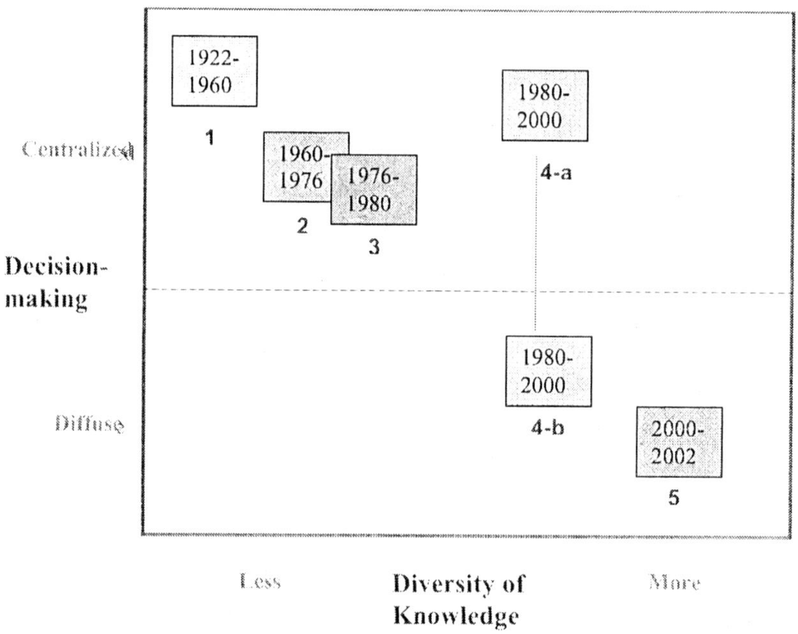

Box 2 in Figure 5 refers to the period of the transitional innovation that produced the first mechanical distributor pump (1960-1976). It reflects both the use of more diverse knowledge and of less centralized decision-making. The first distributor pump design required a more sophisticated body of technical knowledge and a more diverse decision-making process. This begins the period when the injection technology began movement across a threshold from a simple to a complex technology. There is disagreement concerning whether the first distributor pump was simple or complex, but it clearly represents a major design change that was critical to the development of the complex technology represented by the electro-mechanical injection system that would be innovated later. The transition pattern ended with the commercial success of the distributor pump initially used on a large scale in the Volkswagen Golf.

The period represented by Box 2, as indicated in Figure 5, remains characterized by the use of relatively limited knowledge, primarily mechanical knowledge. It also involved relatively centralized decision-making—the design was the product of a small number of people. The mechanical distributor pump, however, represented a more sophisticated and complex technology than the previously dominant in-line pump.

Box 3 in Figure 5 refers to the period after the introduction of the distributor pump in which the incremental innovations were improving both it and the in-line mechanical pump. The improvements during this period would be of major importance to the next transitional innovation, the electro-mechanical design. This period was characterized by an expansion of knowledge, especially an increase in depth; it involved the kind of learning commonly associated with innovations that fit the normal pattern. In sum, there was a substantial increase in the mechanical knowledge utilized, but there was not much increase in the diffusion of the decision-making over what had existed at the time of the initial innovation of the distributor pump.

Boxes 4-a and b in Figure 5 represent the period when the transitional innovations were being carried out that produced the UIS/UPS, VP-44, and Common Rail systems. The period had two phases that were distinguished, among other things, by different electronic controls: solenoid actuators and solenoid valves. The period saw a very great increased in the diversity of knowledge needed. Specifically, it required integrating electronic and electrical knowledge into what quickly became an organizational network of substantial complexity. There was no longer any question about whether the injection technology was simple or complex.

The period reflected in Boxes 4-a and b is particularly interesting in terms of decision-making. Box 4-a reflects the fact that while the diversity of knowledge needed to carry out the transitional innovation experienced a very large increase, at the same time there was a major movement toward centralized decision-making. Active, high level management intervention was necessary to get the self-organizing network that had been carrying out the repeated innovations of the mechanical injection systems to link with and integrate the organizations and people who held the electronic/electrical knowledge. The most significant event in this process occurred when an electrical engineer who had worked for Westinghouse Electric was chosen to lead the Bosch diesel fuel injection division. This major change required centralized decision-making, and that decision-making involved the governing board choosing a leader who was not a "diesel person." The new leadership of then created new organizational relationships—multidisciplinary teams that reported to the leader—that used organizational processes and management actions to integrate electronic expertise into what had been a self-organizing network composed of people with mechanical expertise.

Box 4-b reflects the fact that the innovation of the electro-mechanical systems required both diverse knowledge and diffuse decision-making. The technology that was innovated during this transitional pattern required the full participation of holders of diverse expertise. The complexity of the technology made the existence of a single designer impossible.

Box 5 in Figure 5 represents the present period of innovation characterized by the normal pattern, by repeated incremental innovations that produce specialized configurations of the multiple injection system designs that are in current production. Incremental innovations are being carried out on four different designs. The diversity of knowledge held by the Bosch-centered network makes it possible to rapidly a design that fits the specific needs or desires of diverse customers. Decision-making is very diffuse and it reflects behavior that is consistent with a self-organizing network. Thus, the location of Box 5 in Figure 5 indicates that a very diverse knowledge base and a very diffuse decision-making process currently exist.

5.1. Five Technologies: Knowledge and Decision-Making

The five case studies summarized here demonstrate the variations in diversity of knowledge and differences in decision-making processes that occur between simple and complex technologies, between types of technologies, and between each innovation pattern. Figure 6 illustrates the adaptation paths traced by the knowledge and decision-making of each of the five technologies under investigation.[17]

Figure 6. Evolution of Decision-Making & Diversity of Knowledge in Five Networks

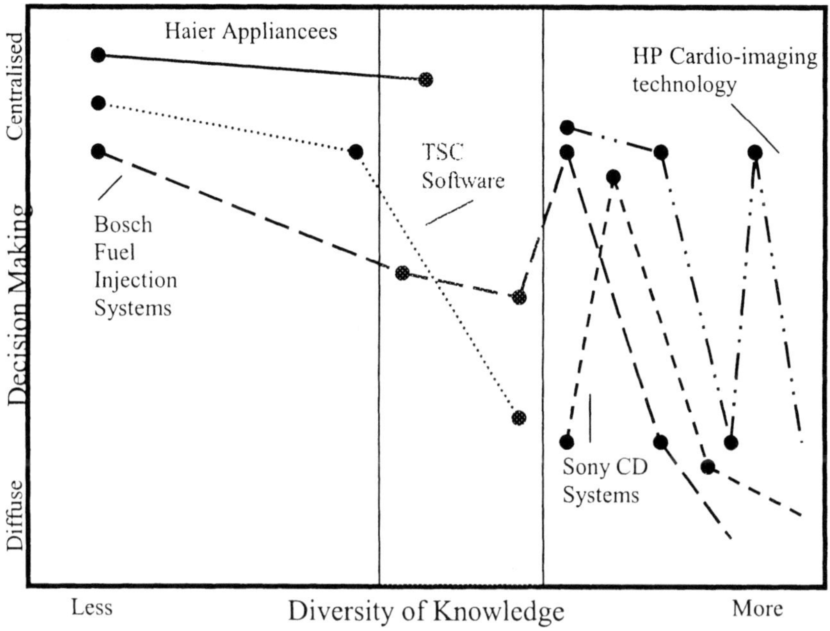

The innovation of the Sony CD is represented by the short-dashed line in Figure 6. The predecessor to the CD—tape systems—represented a fairly diverse body of knowledge and relatively diffuse decision-making processes. However, as the decision to partner with Philips for the design of the CD was made by top Sony management, the decision-making processes were very centralized during the initial period of the transitional innovation pattern. As the transition progressed, there was movement to slightly more diffuse decision-making processes and more diversity of knowledge. With joint Sony-Philips prototype development, during which design decisions were made by consensus, the decision-making became even more diffuse than it was originally, and the diversity of knowledge required (and thus accessed and synthesized) increased significantly.

The HP cardio-imaging technology began the transformational innovation pattern with a diverse body of knowledge directed by very centralized decision-making processes (i.e., the decision by HP management to hire a consultant to explore market opportunities). As the period of transformation advanced, the diversity of needed knowledge increased and the decision-making processes became more diffuse (see the broken line in Figure 6). This is due to the fact that the HP medical products division linked with other HP organizations (e.g., the R&D labs), as well as with members of the end-user community, for prototype design and development. This path of increasing diversity of knowledge and more diffuse decision-making continued in these directions with the subsequent incremental innovations. However, the transitional period of innovation—when HP decided to move to a MRI-based system—once again necessitated very centralized decision-making processes. The creation of the HP-Philips partnership by the companies' management, and their decision to merge networks, is a prime example of the centralized decision-making that occurs when a network is facing a major transition in its innovation trajectory. However, the subsequent joint prototype development and incremental innovations were carried out by a self-organizing network using diffuse decision-making processes to synthesize a very diverse range of tacit and codified knowledge.

The knowledge and decision-making associated with the innovation of Bosch diesel fuel injection technology has already been discussed in detail. This path is shown in Figure 6 by the long-dashed line.

In the case of TCS and its innovation in software services and products, the technology is just beginning a transitional innovation pattern after a long period of incremental (normal) innovation. The path shown in Figure 6 (the dotted line) illustrates its progression from a simple to complex technology. When the technology is simple, the diversity of knowledge is significantly less than that seen with complex technologies. During its period of normal innovation pattern, TCS has increased its diversity of knowledge and has decreased its centralized decision-making accordingly. After the transition to a complex technology, however, a trend of increased diversity of knowledge and more diffuse decision-making will be required.

Haier's innovation of household appliances can be seen as the solid line in Figure 6. As it is concerned solely with simple technology, it represents the least diverse body of knowledge of the technologies examined in our case studies. As it has expanded its line of household appliance products, Haier has increased its diversity of knowledge somewhat. Innovation of the technology has been, and continues to be, directed by a very hierarchical decision-making process. While this is appropriate for simple technology, decision-making processes will need to become more diffuse in order to make a transition from the innovation of simple to complex technologies.

The adaptation paths illustrated in Figure 6 show that each technology is unique in the sense of the combination of knowledge and decision-making processes it requires for innovation. More importantly, however, it shows that even for complex technologies, there are times when centralized decision-making is needed. This is particularly critical as the technologies experience major change—in other words, during the transformational and transitional innovation patterns. It is during these periods that public policy makers and company managers need to be most actively involved. These patterns require leaders to guide and facilitate the innovation process.

During the normal pattern, innovation of complex technologies is facilitated by networks that self-organize to access and synthesize the diverse body of tacit and codified knowledge needed to rapidly carry out incremental innovations. This is especially important because these incremental innovations are the ones that most benefit the innovators financially.

During the normal pattern of innovation of simple technologies, centralized decision-making is possible. Successful centralized decision-making can result from the understanding that comes from extensive codification of the needed knowledge or because the decisions makers are able to have the range of tacit knowledge needed to make informed decisions. As the technologies progress toward the complex threshold, major efforts are usually made to codify existing tacit knowledge. As complexity increases and the rate of innovation increases the codification response becomes less and less adequate. At this point only networks will suffice.

6. THE INFLUENCE OF CULTURE

Culture is an important influence on how and how fast organizations and networks are structured and what processes are used. We have used seven variables throughout our case investigations in an attempt to determine the role that national and company cultures play in facilitating or impeding the innovation of complex technologies. These variables are: ability to trust, comfort with tacit knowledge, comfort with codified knowledge, equality vs. inequality, individualism vs. collectivism, centralized vs. consensus decision-making, and comfort with uncertainty.[18] In the following discussion, due to the boundaries set for this chapter, we refer to these variables briefly as they are particularly relevant.

The three innovation patterns necessitate certain organizational requirements, and subsequently, cultural requirements (see Table 2). The innovation of complex technologies requires the capability to access and synthesize diverse bodies of tacit and codified knowledge that are held in different organizations. Carrying out the synthesis of diverse kinds of knowledge requires organizational networks. Synthesis is a key core capability in the innovation of all complex technologies and it must be viewed as a factor of equal importance to capital and labor.

Table 2: Organizational & Cultural Requirements of Innovation Patterns

Innovation Pattern	Organizational Requirements	Cultural Requirements
Transformational	• Ability to access, create, and synthesize knowledge from sources and with organizational arrangements that are new • Repeated trial and error • Ability to learn from failures	• High levels of trust with those not previously known • Comfort with tacit knowledge • Comfort with codified knowledge • Centralized decision-making • Comfort with uncertainty
Normal	• Ability to self-organize to facilitate rapid learning and unlearning • Problem-solving R&D and non-R&D knowledge • Consensus on next step	• High levels of trust with those previously known • Comfort with tacit knowledge • Comfort with codified knowledge • Consensus decision-making
Transitional	• Ability to integrate separate organizational networks and bodies of knowledge into new network and knowledge base • New routines and heuristics developed by trial and error • Ability to meld cultural differences	• High levels of trust with those previously known and unknown • Comfort with tacit knowledge • Comfort with codified knowledge • Centralized decision-making

Culture—whether national or company—can facilitate or impede the ability of networks to synthesize and thus the ability of networks to successfully carry out the innovation of complex technologies. Where national culture tendencies appear to facilitate innovation, companies can create and maintain cultures that are consistent with that of the national culture. Where national cultural tendencies impede innovation, companies can create and maintain cultures that compensate for the broader norms. Similarly, public policies can be adopted that take into consideration the necessary organizational requirements and cultural requirements. These policies can take advantage of cultural traits conducive to innovation and attempt to mitigate the effects of cultural traits that impede innovation.

The times that require policy changes or company strategy changes are the transformation and transition periods. If a country or company has a culture that supports the use of tacit knowledge and diffuse decision-making and can act quickly on what are effectively consensus decisions, then it will have an advantage similar to the one that Japan has traditionally had with respect to the incremental innovation of complex technologies. For example, Intel has created a company culture that has many of the characteristics of Japanese culture. Intel's company culture emphasizes the importance of giving information and knowledge away quickly, diffuse decision-making, and teamwork. It has fostered a cultural environment of "disagree and commit": once decisions are made, everyone actively supports the decision. In Japan, these kinds of explicit company rules are not required, as the traits of the national culture already support these organizational requirements. The following section will discuss examples of the influence of culture on the innovation of our five technologies.

7. FIVE TECHNOLOGIES, FIVE CULTURES

In the Bosch case, the decision to carry out the transformational innovation—the initial design of a diesel fuel injection system—is consistent with the hierarchical, codified culture of Germany at that time. The technology does not require anything counter to traditional German norms. On the contrary, the technology required craftsmen with narrow specialized technical expertise—both highly regarded in German culture. Once in the normal pattern of innovation, the dominant culture of the company and its organizational network is one of established routines and heuristics that have either engineered around or taken advantage of cultural traits. The normal pattern of innovation of diesel fuel injection systems appears to fit nicely with the German preference for ever greater depth of specialized technical knowledge—both tacit and codified. Significant stress on the organizational system occurred at the second transition point, when the technology transitioned into an electro-mechanical system. It is at this point that Bosch implemented organizational changes in an effort to synthesize new knowledge and to overcome the hierarchy and highly specialized narrow focus of its personnel. During this period of transition, the company was no longer able to organize itself as a symphony orchestra:[19] assembling a number of highly specialized experts led by a "Master" designer. Rather, the transition required a merging and synthesis of expertise in which the interacting linkages and boundaries between disciplines were no longer clear. Thus, the company implemented a new organizational structure that established multi-disciplinary teams.

The Sony innovation trajectory begins with what can be characterized as a self-organizing network: it has a style of operation that utilizes tacit knowledge to synthesize diverse types of knowledge and uses group decision-making. This is highly consistent with Japanese culture more generally. When the transition point comes, and there is a need to link with Philips for purposes of designing a CD system that integrates digital electronics and photonics, however, centralized decision-making was necessary to create the link and to create the requisite speed and urgency. The literature states that Japanese culture generally emphasizes the importance of relationships being mutually beneficial. This trait is obviously advantageous in the innovation of complex technologies, as an organization or network cannot operate in a tacit knowledge environment over a period of time without this attitude. Thus, this cultural characteristic is quite beneficial to the Sony-Philips partnership over the design phase.

Having a design in place, the Japanese focus was on getting the product to market quickly. Sony believed in its tacit knowledge and experience, so it approached future problems that would arise with incremental innovations as something to be dealt with later after the product was on the market. Philips, however, preferred codified knowledge and understanding in the scientific sense. It wanted to ensure that the prototype included all necessary specifications to enable all follow-on incremental innovations.

In the HP cardio-imaging case, the company, in the early period, can be characterized as having a "not invented here" syndrome—a rejection of knowledge and expertise from outside organizations. This was particularly the case when it needed radar knowledge for the transducer. The inability of the HP personnel to recognize the importance of tacit knowledge that was obtainable from other organizations reflected the traditional American cultural preferences for codified knowledge. This substantially slowed down the innovation process, but HP did learn from this failure.

When the MRI transition point approached, HP management decided to join forces with Philips in order to save time. This approach worked with the Philips' culture which was organized into distinct technology areas. Thus, the merging of the networks was easily facilitated. It is important to note that HP was a relative risk taker: it was willing to develop a technology that a consulting company said would sell but for which few cardiologists saw a need. One of the reasons for this risk-favorable organizational culture were the norms instilled by the founders, Hewlett and Packard, they emphasized the importance of developing state-of-the-art technology rather than pursuing short-term profits. By the time that the MRI transition took place, the company had learned the value of tacit knowledge, but it was also operating in an economic policy environment that emphasized short-term profits. Thus, the company culture reflected the broader culture of the American marketplace. Within the normal pattern, the HP centered network was basically self-organizing and used internal learning and rather diffuse decision-making to incrementally innovate.

In the case of TCS, the Indian cultural tradition that emphasizes mathematics, philosophy, and hierarchy works well when the software services and products involve a simple technology. However, as the company transitions to a complex technology and a complex organizational network, it will have to create a company culture that is different than traditional Indian culture. It will need a culture that allows people to act without centralized decision-making and rigid hierarchical organizational structures and one that promotes greater equality. Indian culture is advantageous in the innovation of complex technologies in the sense that it favors a synthesized view of knowledge rather than distinct specialization. TCS instilled a company culture that necessitates the codification of much tacit knowledge when it recognized the need for both tacit and codified knowledge in the innovation of complex technologies.

In Haier's innovation of household appliances, the company has had to create a company culture not entirely consistent with Chinese culture. It has created a very rigid hierarchical management structure with rigid codified personnel and technical performance requirements that are strictly enforced. In order to mitigate the characteristically Chinese culture of creating networks based on familial and friendship ties, Haier has substituted a hierarchical merit-based authoritarian system in order to ensure better efficiency. While it is still concerned with the innovation of simple technologies, Haier can continue to utilize codified knowledge and centralized decision-making. However, in order to transition to complex technology innovation, it will need to utilize consensus decision-making that can handle both codified and tacit knowledge. While Chinese culture is generally comfortable with tacit knowledge, Haier will need to institute strategies and actions that allow linkages with organizations that have the needed knowledge rather than with family or friendship ties. This will be difficult, as Chinese culture emphasizes trust relationships with friends and family. Trust must be earned and experienced before it is assumed.

8. LESSONS

Public policies and organizational strategies designed to facilitate commercial technology innovation benefit from recognizing that very different organizational structures and processes are associated with the success of simple versus complex technologies. The innovation of simple technologies requires a relatively narrow knowledge base composed of both tacit and codified knowledge and centralized decision-making processes are used to guide their innovation. A few key individuals or units within a single organization (or occasionally a very limited number of organizations) possess the total requisite knowledge and capabilities, and make the necessary decisions, in carrying out the innovation of simple technologies. The innovation of complex technologies, however, requires the synthesis of a diverse body of tacit and codified knowledge that resides in many different individuals and types of organizations that link to form a network, and consensus decision-making processes are generally used to guide their innovation. Centralized decision-making is used only at key strategic moments in the innovation of complex technologies. Thus, public policies and organizational strategies that facilitate and encourage cooperation among individuals and organizations are critical to the innovation of complex technologies.

Furthermore, public policies and organizational strategies designed to facilitate the innovation of complex technologies need to reflect the different knowledge acquisition and use processes and the different decision-making associated with transformational, normal, and transitional innovation patterns. The transformational pattern is distinguished by organizational networks with the ability to access, create, and synthesize knowledge from sources and with organizational arrangements that are new. Learning how to synthesize knowledge that is new, and that is from new sources, requires repeated trial and error and is thus heavily dependent on learning from failure. The distinctive characteristic of transitional pattern networks is the ability to integrate previously separate organizational networks and bodies of knowledge into a new, more complex organizational network and body of knowledge held by this new network. New network routines and heuristics are developed by trial and error, and cultural differences pose challenges in the integration of both networks and knowledge.

Thus, public policies and organizational strategies that facilitate the rapid development of trust among the participants in networks are particularly valuable during the transformational and transitional innovation patterns. These patterns represent time of change and uncertainty and require the active participation of company managers and public policy makers to guide and facilitate the innovations. For technologies that gain economic benefits from having a single universal design, public policies and organizational strategies that facilitate non-market choices of redesign appear to be beneficial.

The normal pattern of innovation is distinguished by networks with the ability to self-organize in order to facilitate internal learning necessary to carry out incremental innovation of existing technology designs. A common characteristic of rapid self-organization is the capacity to discard knowledge that is not directly relevant to carrying out the next incremental innovation. (However, this capacity to rapidly discard knowledge that is not immediately relevant is a major barrier to the adaptations necessary to carry out transitional innovations.) Research and development are rarely the most important, and are often not significant, contributors to the knowledge necessary for normal pattern innovation. Normal pattern innovation is guided by internal, often tacit, knowledge and consensus decision-making concerning the direction of innovation. Normal pattern innovation requires less management and public policy attention than the transformational and transitional patterns. This is important to note because it is normal pattern innovation that financially benefits companies most. Economic success for complex technologies generally flows from organizational networks that have a capacity for rapid self-organization. The capability to carry out repeated incremental innovation ahead of in parallel with competitors is essential to success in the market.

The capacity to carry out the synthesis of diverse kinds of knowledge can only be developed by organizational networks. Synthesis is a key core capability in the innovation of all complex technologies and must be viewed by public policy makers and company managers as a factor of equal importance to capital, R&D, labor, or intellectual property protection. At the heart of the capacity of organizational networks to carry out rapid and repeated syntheses is the development of network routines and heuristics. As the process of synthesis is something that networks learn how to do, but never understand, important routines and heuristics will always be network-held tacit knowledge.

Finally, our case studies suggest that a very large portion of the conceptual framework that we have traditionally used to think about and debate public policy and management aimed at facilitating commercial technological innovation is inappropriate for complex technologies. This is important because complex technologies are the most valuable in the world market. The innovation of complex technologies mandates organizational networks for the many reasons discussed throughout this chapter. Consensus decision-making with strategic use of centralized direction should be the norm, rather than the occasional mode of operation. Traditional anti-trust positions impede rather than facilitate the innovation of complex technologies. The emphasis placed on traditional R&D and codified knowledge slow the innovation of complex technologies, as tacit knowledge and learning from repeated trial and error are equally, if not more, valuable.

Chapter Ten
COMPLEX SYSTEMS AND COLLECTIVE ADOPTION:
The Role of Networks and Partnerships as an Endogenous Mechanism to Reduce Dynamic Transaction Costs in the Context of Systemic Innovations

Dominique Foray
OECD

1. INTRODUCTION

A systemic innovation is one whose value, or private and social "return on investment", depends on its adoption by a number of economic agents: for example, the adoption of the experimental method by the scientific community in the 19th century,[1] quality standards, new inventory-management practices, and electronic data-interchange networks. Note that all these innovations facilitate either intellectual or logistical and physical coordination. In each case, the value of the innovation is therefore highly dependent in its collective adoption.

In a complex system, the collective adoption of an innovation can be extremely costly; this fact is expressed by the concept of "dynamic transaction costs". Dynamic transaction costs (DTCs) are the costs of persuading, negotiating, coordinating and teaching outside suppliers, clients or even rivals manufacturing the same kind of products (Langlois, 1992). When a large corporation wishes to adopt a just-in-time order and delivery system, it must have the support of all its suppliers, help them transform their organizational system, and so on. These are DTCs.

These costs can be so high that they actually hinder innovation (the collective adoption of a systemic innovation) and, in some cases, result in vertical integration (when the innovation concerned is particularly significant).

DTCs constitute a plausible rationale for vertical integration, especially in a context of frequent technological and organizational change. Silver (1984) contended that, rather than devoting effort to catching the attention of suppliers and distributors and converting them to new methods, the innovating firm may find it cheaper to acquire the capabilities necessary to produce its own inputs. In other words, instead of incurring high DTCs, integrating multiple production stages may represent a superior organizational structure for coordinating systemic changes. When developing a mass-production system, for example, Henry Ford and his engineers were forced to develop certain special-purpose machinery themselves, as it would have cost more to have it made by outside suppliers (Langlois and Robertson, 1989).

The "flip side" of DTCs is constituted by dynamic governance costs: it may prove expensive to transform internal capabilities in order to adapt them to new standards or technologies when suitable capabilities are available on the market. For example, a company specializing in mechanics will not necessarily integrate electronics capabilities and skills if these already exist elsewhere.

To establish a link with the problem of networks, alliances and partnerships in innovation, DTCs must be endogenized. DTCs can be reduced, depending on the characteristics of the networks and partnerships of the industrial sector concerned.

We would argue that the historical importance and development of networks plays a role in reducing DTCs in an environment of systemic change, and that this, in turn, allows innovation to spread in a concerted manner. We therefore define a network as an institutional mechanism aimed at lowering DTCs in a context of systemic innovation. This definition will be illustrated by means of three case studies representing increasingly complex systems based on the varying centralized nature of decisions on innovation.

2. ABILITY OF A LARGE CORPORATION TO COORDINATE SYSTEMIC CHANGE

Here, the degree of complexity is still low, as the system is fairly centralized —for example, when a large corporation decides to have its subcontractors adopt an electronic data interchange network (David and Foray, 1994).

However, DTCs can be high if suppliers are not prepared (especially from an organizational point of view) for systemic change. In each case, the early establishment of a partnership between the initiating firm and its suppliers will make it possible to reduce DTCs endogenously.

For example, Kelley and Arora (1996) present the case of a group of leading companies (Xerox, Motorola, Kodak, Texas Instrument, Digital, Chrysler) that created a consortium for supplier training and established regional training centers dealing with quality control and cost-effective methods for small suppliers. Asheim and Isaksen (2000) discuss a policy instrument called "knowledge intensive industrial clustering", developed in Limburg, the Netherlands, which aimed at supporting policy for upgrading a big company's network of suppliers (in this case, Océ, a multinational copier producer). Groups of suppliers can also establish centers for training users and distributing technical information in order to gradually introduce new equipment to downstream industrial firms.

The initiation of partnerships by the large company aims at decreasing DTCs in a turbulent environment.

3. ABILITY OF A GROUP OF FIRMS TO FORM A COALITION TO SOLVE COLLECTIVE-ACTION PROBLEMS IN A DECENTRALIZED SYSTEM

Here the situation is more complex, because the relational system is decentralized. In the case of a group of companies in the same industry trying to agree on implementing technological standards, for example, DTCs are high, as the adoption of the innovation in question represents a thorny collective-action problem: since the innovation is systemic (and the value dependent on the scope of its adoption), each firm waits for the other to take the first step before committing itself, which creates an excess of inertia on the market. Such excesses constitute a market failure. In these circumstances, formal institutions such as industrial associations and engineering societies can play a decisive role in coordinating collective action, reducing DTCs to the point that the innovation is eventually adopted.

A fascinating historical example of this phenomenon is the U.S. auto industry's intercompany technical standardization in the early 1900s (Thomson, 1954). The standardization of a certain number of parts and components was needed to deliver a specialized service, achieve economies of scale and specialization, and reduce mutual dependence between suppliers and manufacturers. Although each company was convinced of the private and social benefits of standardization, however, the phenomena of excess inertia and market failure constituted major impediments to the process. The role of the Society of Automotive Engineers (SAE) was critical to the development of the standardization process.

A more recent illustration is the insurance market in London, as described by Rae (1999):

> All the firms were facing similar problems of cost and complexity (...) They identified a pressing need to do some or all the following:
>
> reduce the costs of storing and sharing information;
>
> reduce duplication of work by different firms;
>
> add more structure to the information, so that clerical tasks can be automated;
>
> add structure to support the use of quantitative techniques.
>
> To meet these challenges, opinion coalesced around a market

network, open to all, with market standards for data and processing rules.

The London Insurance Market was facing a collective action problem. In this case, traditions of knowledge sharing and the role of industrial partnerships as forums for the exchange and existence of interpersonal networks, are factors contributing substantially to the minimization of the DTCs, and as such facilitate concerted actions for the collective adoption of the innovation.

In all these cases, partnerships play just as critical a role as an endogenous mechanism to decrease DTCs and facilitate the collective adoption of systemic change.

4. THE DUAL COORDINATION PROBLEM : WHEN AN INNOVATION DEVELOPS AT THE PERIPHERY OF A CENTRALIZED SYSTEM

The system here is even more complex, insofar as, even though it is still the centralized system found in Figure 1, innovation and initiative come from the periphery. An example of this type of innovation deserves mention (Nantua *et al.*, 1999).

The innovation in question came from a junior bearing manufacturer, and involved the principle of electronic data capture. The great idea was to use the wheel bearing as an active sensor, enabling the collection of a lot of data. The technical device helped solve a number of ABS-brake system problems.

For this small concern, the challenge represented by innovation development and widespread application was enormous, forcing the company to adopt a triple-faceted form of alliance discussed in depth by Ed Steinmueller (this volume). We can identify these three forms on Figure 1:

Figure 1. Innovations in Complex Systems:

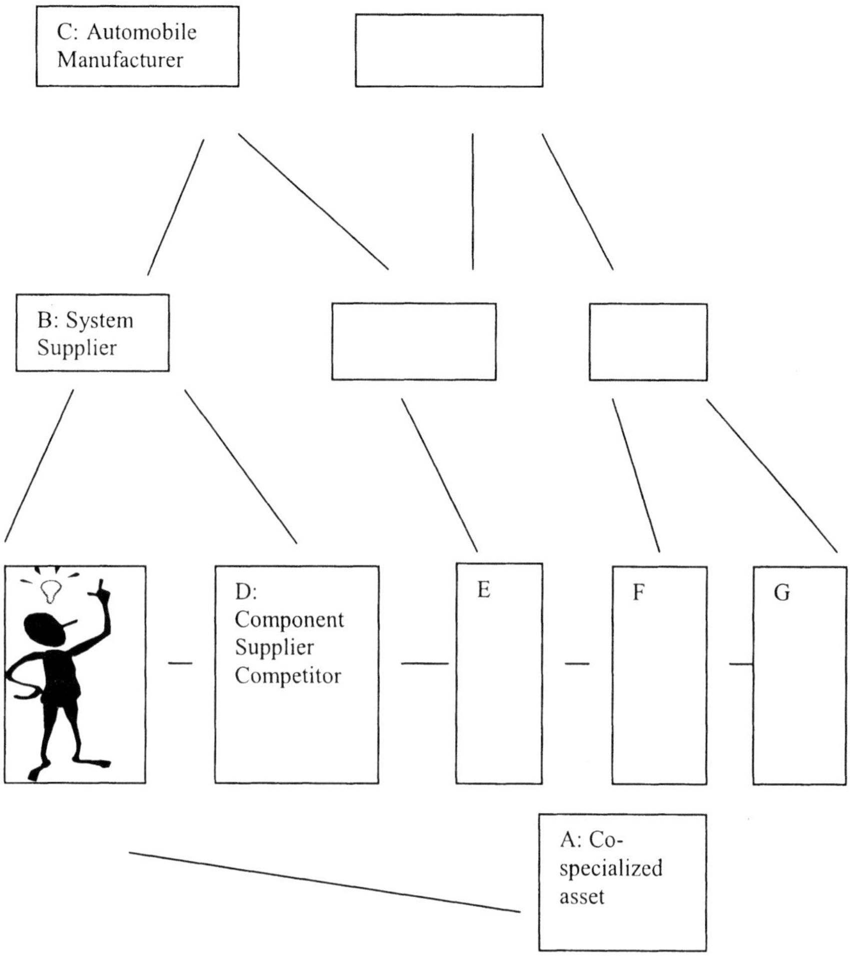

- An invention-based alliance: The company specialized in mechanics, and electronic developments obliged it to enter into agreements with specialists in the field (company A on figure 1). This involved the creation of co-specialized assets (as described by Teece, 1986), a process that ended with an innovation "that works" (at least in theory);

- A design-based alliance: Soon the firm started working with leading equipment suppliers and automobile manufacturers (companies B and C) to define the new component's interfaces, so as to insert it into the subsystem. This phase ended with an effective, standardized component;

- An integration-based alliance: One obstacle remained, however. By innovating, the small company became unique in the eyes of the major manufacturers, and this made the latter vulnerable and dependent, encouraging them to wait until the innovation was adopted by all subcontractors. The small company therefore established a coalition by persuading its competitors (companies D, E, F, and G) to adopt the innovation as well. To survive, that innovation had to become the industry standard.

Overall, the innovation process was extremely lengthy, lasting from 1984 to 1997. It was punctuated by dramatic strategic revisions, for example with respect to intellectual property. The integration phase made it necessary to sacrifice patents taken out at the end of the invention phase, and "open up" the technology so it could become standard.

This account can be understood in terms of DTCs' and networks' and partnerships' abilities to reduce costs. In phase one—invention and alliance with electronics experts—dynamic governance costs (the costs for the mechanics firm to integrate electronics capabilities) were much too high in relation to the immediate availability of those capabilities on the market.

In phase two—design and partnership with downstream firms—DTCs were high, and necessitated relatively solidified forms of coordination (consortium bringing together the different stakeholders in the innovation-design process).

In the third and final phase—which involved integration and the establishment of a coalition to establish a standard—DTCs skyrocketed. How could competitors be persuaded to join in, to adopt a technology invented elsewhere and share the risks of implementing the innovation? The fact that an alliance or network of subcontractors did not already exist made these costs even higher.

As it turns out, the story had a happy ending: once the innovating firm's strongest competitor decided come on board, the rest was history. The innovation was on its way to becoming a world standard. The DCTs associated with each phase—invention, design, and integration—were lowered by means of specific partnerships aimed successively at producing the invention, improving the design, and creating a standard.

5. DTC DEVELOPMENT TRENDS IN THE NEW ECONOMY

There is an extensive amount of literature on the fact that the knowledge-based economy is one characterized by an acceleration in innovation. The expressions "restless capitalism" and "learning economy" both try to describe the same phenomenon: Significantly increasing investment in innovation (not least in R&D) has sent the numbers of innovations appearing soaring, as evidenced not only by the volume of patents requested and approved (OECD, 1999), but also by the proliferation of new varieties of goods and services that has marked the trend toward "mass customization". Meanwhile, the "need to innovate" is growing stronger as innovation comes closer to being the sole means to survive and prosper in highly competitive and globalized economies. It is not easy to distinguish between absolute novelties ("under the sun") and innovations that are new only to the companies that adopt them, or more complex adaptations of existing products or ideas to a new market. The fact remains that companies and society in general are spending more time and energy on producing and adjusting to change.

If we consider both this trend and the one that characterizes increasingly complex technological systems, it seems likely that a DTC explosion will occur. The costs of the collective adoption of innovation will likely get higher and higher. This has two implications:

- First, partnerships and networks will become increasingly important as an endogenous mechanism for reducing DTCs; and

- Second, more and more technological systems will be designed around a modular structure (Baldwin and Clark, 1997), which establishes a distinction between the external and the internal design of a component or subsystem. Modular construction establishes visible design rules, especially those governing the external interfaces between modules in order to ensure integration and interoperability, while internal module design remains in the hands of the innovating firm. By reducing coordination to visible rules, modularity aims at lowering DTCs in a context of rapid systemic change.

Current developments in terms of innovation intensity would thus seem to be leading to an increasingly important role for networks and modular organization as dual coordinating principles for reducing DTCs.

6. CONCLUSION

Our analysis has revolved around the concept of DTCs, which corresponds to the cost of the collective adoption of a systemic innovation. The extent of these costs is a plausible explanation for vertical integration and the establishment of networks and partnerships. Our case studies show that such institutional arrangements constitute endogenous mechanisms that are often effective in reducing DTCs.

The measurement of DTCs is unquestionably a major challenge: gaining an understanding of their quantitative evolution would be extremely useful in establishing the importance of networks and partnerships in a context of systemic innovation.

An evaluation of how much time engineers in large corporations spend transferring technological capabilities or organizational practices among suppliers and subcontractors would constitute a useful indicator. Another indicator would be the budgets allocated by companies for the creation of forums and partnerships intended to educate and familiarize customers, subcontractors and suppliers on and with successive waves of systemic change. Although this kind of measurement program is still too "blue sky", the importance of the DTC concept, as well as the growth of empirical studies on the measurement of transaction costs, would seem to justify the allocation of significant resources for such a program.

Chapter Eleven

INDUSTRY LIFE CYCLE, KNOWLEDGE GENERATION AND TECHNOLOGICAL NETWORKS

Lionel Nesta and Vincent Mangematin
INRA, Université Pierre Mendès-France

1. INTRODUCTION

Industry Life Cycle theories have launched a vigorous research program studying the forces governing the evolution of industrial structures (Abernathy and Utterback, 1978; Utterback and Abernathy, 1975). As such, they challenge the traditional approach of industrial organization by underlying the intrinsically changing nature of industries. Evidently, industry structures evolve on the basis of market selection. However, as technologies and competencies, mobilized in a given production chain do change, firms ought to devote additional resources to learning and eventually to mastering new knowledge. Less efficient actors are not only those that fail in implementing their own production function in the short run. They may also be those that fail in modifying it in the long run (March, 1991).

Fundamentally, life cycle theories exhibit two major phases in industry development, each being related to the state of the technological paradigm at stake (Dosi, 1982; Tushman and Anderson, 1986). While the first phase is characterized by radical and rapid technical change, the second reveals some sort of technological consolidation and stabilization around a dominant design (Anderson and Tushman, 1990). The rising recognition of the role of knowledge in industries may well be associated with the corresponding emergence of the so-called knowledge base literature (Afuah and Utterback, 1997). Briefly, this fruitful research avenue emphasizes the fact that firms must engage in a wide variety of learning processes (Malerba, 1992) not only to improve their on-going production process or innovate with new products, but also to cope with rapidly changing technological landscapes. In fact, this has quite important implications in knowledge intensive industries, for mastering a wide range of scientific and technical knowledge often proves unachievable for one isolated organization. Instead, firms would rely on extra-organizational arrangements, supporting the view that firms are not isolated islands of production (Richardson, 1972). Empirical facts strongly support the view that firms are embedded in complex networks of alliances and collaborations. Behind the explanation of why firms establish ties lies the assumption that firms naturally benefit from the complementarity of their partners' assets and competencies (Hite and Hesterly, 2001). Several studies tend to show that profoundly interacting firms tend to achieve higher levels of performance (Powell, *et al.* 1996). In fact, it is not the mere number of alliances that prove important. Rather, the firm's central position and the density of its network boost their innovative performance, their market value and their operating income (Afuah, and Utterback, 1997; Baum, *et al.*, 2000; Stuart, 2000; Stuart, *et al.*, 1999).

While the benefits drawn from networking are gradually recognized in the economic literature, much less is known regarding the temporary or permanent structure, or even presence, of networks. Put differently, ought networks to disappear with the stabilization of industry structure, or do they express a new form of industrial organization on its own? In the absence of clear empirical evidence, two substitutive views are confronted. Tenants of the former view posit that inter-organizational alliances represent a solution for firms to cope with the instability of the technological environment. As technologies settle down, the need for external alliances becomes more blurred. Arms length contracts may well prove more efficient economically, and the underlying industrial organization concentrates around the traditional distinction between markets and hierarchies. Eventually, networks ought to naturally evaporate. Tenants of the continuity of networks hold the opposite assumption that the very nature of technologies implies systematic ties, alliances and collaborations amongst various types of actors. Here, networks constitute a substantive form of organizational structure by itself, complementary to the traditional distinction between markets and hierarchies (Orsenigo, *et al.*, 2001; Orsenigo *et al.*, 1998). In a way, the two contradictory propositions differ in the assumption regarding the changing nature of technology. While the former pictures technology as following a specified path of non-linear but expected consolidation, the latter states that technology on its own has become intrinsically complex, profoundly rooted in the so-called knowledge based economy.

Thus, a natural way to explore both views is to tackle the evolution of a given technology and explore the actual network structure of the main actors involved in it. The objective of this chapter is hence to analyze the links between the rhythm of knowledge creation, the types of actors involved in knowledge generation and the structure of networks of technological collaborations. The analysis, results, and conclusions shall be concerned with intra-paradigmatic knowledge dynamics exclusively. We define the latter in terms of rate, novelty, and conditions of knowledge combination within one of the most important technological breakthroughs of the twentieth century: biotechnology. Operational measures are derived using patent statistics applied for between 1975 and 1998. The research is organized as follows.

Section 2 reviews the theoretical and empirical literature as a background for the elaboration of four propositions of research. In section 3, we expose the data and method used in our fieldwork. The statistical results are commented in section 4. Two main results are presented. First, the number of patent applications is continuously increasing, though the number of radical innovations (new technological arrangements) reaches rapidly a steady state. Second, both the number of actors and the number of collaborations follow a bell shape, thus seeming to positively correlate with the rate of technological change. Section 5 discusses and concludes the chapter.

2. LITERATURE REVIEW AND HYPOTHESES DEVELOPMENT

2.1. From Exploration to Exploitation

During the past three decades, propositions concerning technological evolution have been advanced to provide a plausible explanation of the sources of the so-called industrial life cycle. In this vein, concepts like dominant design (Utterback and Abernathy, 1975), technological trajectories and regimes (Nelson, and Winter, 1982), technological guideposts (Sahal, 1981) and technological paradigms (Dosi, 1982) illustrate the idea that processes of knowledge generation and exploitation do not follow linear paths. Importantly, such studies of technical change insist on the idea that the accumulation of knowledge follows different, cyclical phases rather than the mere linear process of knowledge accumulation.

Perhaps the major feature of technological discontinuities is to introduce both new questions and their associated set of techniques that challenge pre-existing methods of doing science and/or using techniques. Two research traditions, or paradigms, can therefore compete on the basis of their unit of analysis, the associated set of techniques, leading to competing explanation of similar phenomena. Technological discontinuities occur at specific stages of knowledge development. Radical change is expressed when such discontinuities challenge and partially invalid traditional paradigms. But within the new paradigm, different methods may well compete with one another. In other words, time for knowledge exploration and refinement is still needed in order to set standardized procedures of knowledge exploitation.

The above remarks suggest that intra-paradigmatic knowledge development follows at least two different and probably successive phases. The first one is characterized by fast knowledge creation. New procedures are explored. New technological combinations are frequently introduced in the technological system as to determine their respective effectiveness. But as new knowledge piles up, the marginal improvement is expected to decrease overtime. That is, new technologies tend to only marginally improve the existing intra-paradigmatic stock of knowledge. Here, the technological system moves onto the second stage, i.e. the establishment of a dominant design. A design is said to be dominant when its utilization prevails over that of other knowledge (Anderson and Tushman, 1990). Importantly, the establishment of a dominant design refers to the economic production and commercialization of technologies. Nelson and Winter (1982) define a technological regime as a particular combination of fundamental properties of technologies: opportunity and appropriability conditions, degrees of cumulativeness of technological knowledge, and characteristics of the relevant knowledge in use. Importantly, these properties go hand in hand with the development of the scientific sphere of knowledge development. It posits that the technological regime follows a similar dynamic as knowledge creation reaches an asymptote. Thus we advance our first propostion.

P1: As the industry matures, the rhythm of knowledge creation is slowing down.

Taken as it stands, the above proposition must be linked to more economic phenomena. In fact, the important idea here is that technical change is directly interacting with industrial structures. New firms may well be created and might subsequently grow on the basis of their distinctive technological skills, gaining access to preferable positions on new market segments. By way of consequence, the emergence of a new paradigm may potentially destroy the traditional industrial barriers to entry, thus representing a threat to incumbents. Therefore, the development of knowledge, or technologies, should not be thought of as given for, or provided to, incumbents of a particular industry. Rather, firms must devote efforts in implementing learning strategies so as to keep up to date with relevant knowledge and techniques. Likewise, the establishment of a new dominant design calls for knowledge exploitation more than knowledge exploration. As March (1991) pointed out, firms aiming at diversifying their knowledge base might be better fitted in adapting in and to an unstable environment. Yet with the stabilization of the technological environment, exploitation strategies, i.e. the ongoing use of the firm's knowledge base (Vermeulen and Barkema, 2001) helps the organization to refine its routines and allows it to recoup initial investments and to become profitable. Thus during the exploitation phase, firms focus on the knowledge that contributes most to its success, and filters out knowledge and routines that are less successful.

In this line of thought, one can also say that the firms' knowledge base reflects to some extend the state of the technological environment in which firms are embedded (Afuah and Utterback, 1997). When a dominant design appears, the choice between the different hypotheses explored reduces drastically. Thus firms might prefer to concentrate on their internal capabilities and improvements of existing capabilities, stemming from the integration of bodies of knowledge whose services has proven more profitable. Likewise, Malerba and Orsenigo (1997) identify two patterns of innovation, namely Schumpeter Mark I and Schumpeter Mark II. Schumpeter Mark I is characterized by the creative destruction with technological ease of entry. In this setting, entrepreneurs and new firms play a major role in renewing industrial knowledge and by reaching higher innovative performances. Conversely, Schumpeter Mark II is characterized by creative accumulation with the prevalence of large established firms and the presence of relevant barriers to entry for new innovators. Thus the authors show that both the industry structure and the nature of innovative activities are related to specific ways of accumulating knowledge. While Schumpeter Mark I relates to the first phase of intra-paradigmatic knowledge development, Schumpeter Mark II relates more to the second phase, namely that of consolidation. Importantly, the latter leads to an opposite move on the industry structure. Firms' growth, be it in terms of size or by means of acquisition or mergers, leads to an industrial shakeout (Klepper, 1997). That is, a substantial number of firms exit the industry, while survivors take stock of their past successes. This leads us to our second proposition:

P2: As the sector matures, the number of actors tends to diminish

Proposition 1 and 2 stem from the stylized facts derived from both theories and empirical analysis of the industry life cycle. Such theories provide a convincing account of how and why industries do evolve overtime, underlying the role of industrial knowledge and competencies and revealing the changing number of actors involved in the very course of market competition. Nevertheless, they tend to downplay the role of other actors that might well prove central to relevant technological trajectories. Our main contention here is that academia is often the source of technical change, stemming from scientific developments of peculiar disciplines. As such, not only the notion of technological paradigms is similar to that of scientific paradigm (Kuhn, 1962), but both do not represent independent spheres of actors (Murray, 2001). Rather, both are closely intertwined and are linked on the basis of their scientific and technical content.

Recently, there has been a substantial rise in the number of contributions revealing the economic role of academia (Brooks, 1994, Gibbons *et al.*, 1994, Mansfield, 1995). Such contributions state that it is nowadays difficult to establish a clear frontier between academia and industries. The underlying reasons are many. First, the intensification of knowledge intensive activities has contested the traditional separation between fundamental and applied knowledge (Nelson, 1959). Firms mobilize both fundamental and applied knowledge in the very conduct of their productive activities. Thus, both the sequential and institutional distinction between an academic sphere initially concentrating on the creation of fundamental knowledge and an economic sphere subsequently exploiting applied knowledge can no longer hold. Second, public and private research laboratories have developed tight links irrespective of the industry. The past two decades have witnessed a considerable upsurge in research and development (R&D) collaborations throughout market economies that transcend the traditional partition between both spheres. A third important consequence of such a view is the rejection of sequential phases of knowledge development. Rather, both academia and firms play a central role in all knowledge development, from fundamental to applied knowledge and vice versa.

We do not intend to contest such well-accepted phenomena. We do also acknowledge the fundamental switch that has occurred with the move towards knowledge intensive activities and/or societies. Whether academia or firms are the sources of technological breakthroughs does not lie in our line of inquiry. But following Dasgupta and David (1994), the recognition of the links between public and private laboratories should not lead to the denial of their differences with respect to their cultural background, their incentive structures, their rewards systems, etc. It follows that academic research units are supposed to be more active in knowledge creation than in innovation per se. Likewise, we might conjecture that the growing economic exploitation of given bodies of knowledge should lead to a higher involvement of private laboratories in knowledge development, that is, to a consistent rise in the number of firms involved in the economic exploitation of knowledge. Thus we might expect that the relative share of private laboratories in patenting rise overtime. We posit:

P3: As the sector matures, the role of academic research in patenting tends to diminish

The above propositions have been advanced at the theoretical level only. Actually, we shall be concerned with the analysis of the biotechnology industry, for the moment leaving apart its intrinsically heterogeneous nature. During the past thirty years, the life science industry (the seed, the agro-food, the chemical and mostly the pharmaceutical industry) has gradually awakened to the potential of biotechnology and has subsequently attempted to integrate and to master it. Of particular importance has been the emergence of new fundamental knowledge derived from molecular biology and genetic engineering. The rise and development of biotechnology was based on the project to calculate the structure and properties of biological macromolecules and cells from the physical interactions of their elementary components. Advances in this direction greatly improved the ability of researchers to rise the predictive power of the R&D activities based on the calculation of the functional properties of given molecules. Moreover, the convergence of molecular biology with information technology has brought powerful computational capabilities to biologists, allowing the storage and exploitation of very large-scale information sets. For instance in pharmaceuticals, new techniques such as high-throughput DNA sequencing, functional genomics, bioinformatics and proteomics are becoming fundamental tools in modeling the structure of nucleic acids and proteins, thereby providing pharmaceutical researchers with new rational tools for drug development.

Today, it is clear that the knowledge of molecular biology and genetic engineering represented a discontinuity with respect to the knowledge previously used in life sciences (Arora and Gambardella, 1994, Oliver, 1999). The emergence of the industry was based on the exploration of large numbers of scientific hypotheses. A growing number of actors (academic researchers, large firms and start ups) have been involved in the exploration of hypotheses. Perhaps the most striking feature related to the development of biotechnology has been the high level of firm creation since the seventies. To date, more than 3,000 dedicated biotechnology firms (DBFs) have been created throughout the world, most of which being located in the American and in the European continents (Ernst&Young, 1998; 2000). In our second proposition, we have argued that an industry shakeout should eventually be expected. Thus we now turn to another puzzling characteristic related to biotechnology, that is, inter-organizational collaborations and alliances.

Actually, biotechnology has traditionally been based on close ties between the university and the industry (Kenney, 1986). This has many justifications. First, the very nature of biotechnological knowledge blurs the previously-mentioned frontier between fundamental and applied knowledge. Discoveries in the former may boost developments in the latter while new applications may well lead to new scientific breakthroughs. As a consequence, public and private laboratories have extensively relied on inter-organizational arrangements as a method to share differentiated yet complementary knowledge. A second and subtler consequence of the peculiar nature of biotechnology knowledge is that university researchers have been in the best position to exploit the potentiality of biotechnology. Indeed, many DBFs were university spin-offs, being under the scientific direction of university professors, and being actually managed by former academic scientists. Altogether, it is clear that university-industry relationships in biotechnology have been the rule rather than the exception.

Inter-institutional collaborations provide an extensive account of patterns of collaborations. Yet they do not exhaust it. Alliances with potential clients and other firms involved in biotechnology are needed as well (Oliver, 2001; Powell, *et al.*, 1996) generating new patterns of learning. Co-operations take different forms, from commercial, research or production agreements to joint ventures and mergers or acquisitions. Recent research on alliances and networks has stressed the value of inter-organizational relationship for assessing resources and creating competitive advantages (Dyer and Singh, 1998). The expected benefits of alliances and networks are the enrichment of the flow of knowledge amongst partners (Kogut, 1988), the access to complementary assets (Pisano, 1990) as well as access to external legitimacy and status (Baum and Oliver, 1991; Liebeskind *et al.*, 1996). Notably, Baum *et al.* (2000) show that a start-up's initial performance increases with the size of alliance network at founding and with the efficiency of its alliance network at founding. The authors explain how the roles of alliance vary amongst firm age and size. At founding, the variation in alliance and network composition produces significant differences in their performance, supporting the idea that liabilities of newness and smallness result, to a large extent, from a lack of access to resources and stable exchange relationships.

Disappointingly, such contributions do not provide a substantial account of the changing nature of the firms' patterns of networks over time. Our contention is that the exploitation of few promising scientific results must eventually lead to corresponding changes in the structure of networks. At the level of the firm, one may well conjecture that the network structure of a given firm changes together with its size, its age, or more globally its growth. Indeed, several qualitative case studies (McNamara and Baden-Fuller, 1999; Steier and Greenwood, 2000) show that the composition of collaborative networks in biotechnology changes as firms grow. Commercial and financial alliances become more vital while alliances with other firms or public laboratories gradually decrease. At the level of the industry, we might equally speculate that the changing nature of the technological environment leads to different patterns of collaboration. Notably, since firms seek to appropriate knowledge and translate it into private rents, be it higher productive processes or new commercial products, the role of collaborations in innovative activities should eventually diminish as compared to earlier research phases. We propose:

> P4: Collaborations and co-patenting between public and private or amongst private organizations tend to diminish as the industry matures.

3. DATA AND METHODS

Our study is based on Derwent Biotechnology Abstracts (DBA). DBA is a database covering all biotechnology patents applications since 1981. Today, more than 96,284 patent applications are reported in DBA, from 1965 to 2001 and covering 40 intellectual property authorities. Because two years are needed for inventory purposes, the curve drops tremendously after 1998. Therefore, the analysis shall exclusively be concerned with prior patents applications.

DBA provides a variety of information about patent applications: the name of inventors; the patent affiliate; the date of applications; extensive summaries; key words; etc. Computer-expensive manipulations were necessary in order to translate the initial DBA information set into a workable database format. Particular attention was paid to the patent's year of application and the name of the applying organization. Predominantly demanding was the treatment of names of organizations. Discarding missing information condenses the database to 74,905 patent applications emanating from 3,483 different names. Recurrent name misspelling was treated at length, while fusions and mergers occurring amongst firms were further taken into account. In total, 1,469 organizations were identified. Names provide valuable information about the institutional type of organizations. For instance, universities and institutes gave rise to multiple abbreviations that allowed us to easily differentiate public research organizations and private organizations, be they large firms or DBFs. A final ad hoc checking of names was performed using mainly international biotechnology directories and additional Internet resources, allowing us to geographically locate our sample of organizations. Altogether, 898 private organizations and 571 public research organizations were identified, accounting for 78% of the overall number of patent applications recorded in the original database.

Here, it is worth stressing a number of observations. First, one should be aware of the fact that we are concentrating on a peculiar database whose coverage remains somewhat conditioned to the DBA definition of biotechnology. Whether a given patent application falls into DBA's boundaries of biotechnology is thus restricted to some subjective criteria. The important point here is that when actually analyzing DBA patent applications, we are all the more concerned with a subset of an unknown exhaustive population of biotechnology patent applications. However, we are confident in the fact that the DBA coverage focuses on the core of biotechnology, while more technologically distant patent applications are discarded. Second, the modalities of actor selection imply that only relevant actors are included in the analysis. This suggests that when analyzing the evolution of the number of biotechnology actors, we are actually speaking of relevant, important, actors, discarding marginal yet potentially essential other phenomena. Third, one should also keep in mind that DBA is a worldwide database, thus gathering different national systems of innovation together with several approval authorities. One of the most vivid manifestations of it lies in the relative share of bounded geographic areas in the total number of patent applications. Particularly, North America accounts for more than one third of all recorded patent applications (29,009), as Japan does (27,927), while the European continent hardly amounts to one fifth (12,932).[1] While our own knowledge of biotechnology conforms to the observation of the American share, we are more dubious about the Japanese share.

Certainly, the previous remarks call for caution when commenting on our results. The fact that only the core of biotechnology is analyzed here naturally outplays additional events that, though marginal, may yield further discontinuities in the intra-paradigmatic development of biotechnological knowledge. Though aware of that, our motivation in using DBA is above all practical, conforming to a substantial number of other authors (Archibugi, 1992; Griliches, 1979; 1990; Pavitt, 1988). DBA represents a unique database providing us with a chronological and systematized information set concerning both the rate of knowledge creation and the number of actors involved. Besides, DBA allows us to further describe and analyze the relationships between technical change and the underlying network structure.

Concerning the analysis of technical change, we shall essentially give attention to the technological content of patent applications. Of course, there exists a multiplicity of methods by which one could exploit such a large dataset (see notably: Frenken, 1998; Saviotti, 1988). While agreeing with the relevance of such methods, we focus on the technological combination of patent applications so as to assess novelty in the so-called technological space. Indeed, DBA characterizes each patent along thirty biotechnology-dedicated classes (see Appendix 1). Theses classes were converted into a vector of 30 technology classes, taking value 1 if technology i occurs in patent P, 0 otherwise. For example, if technologies A1 and A2 occur within patent P, P can be described by the 30 dimensional vector I = {1,1,0...0}. It follows that all patents may be described along a thirty-digit string that we call a technological combination or arrangement. Within one patent, a maximum of 6 technologies may be combined, yielding a maximum number of about 768,000 possible arrangements. Since only 2,525 arrangements have been identified in DBA, the construction of the technological space exhibits far-from-random explorative learning strategies.

Figure 1. Sets and Subsets in DBA. By Date, Identified Actors, and Co-Patent applications.

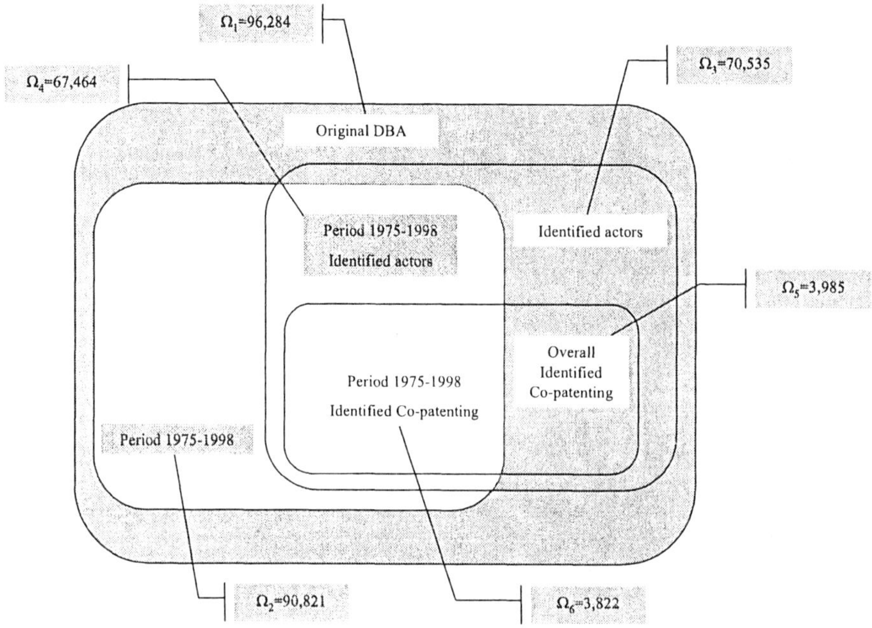

Turning to the description of network structures, co-patent applications enhance the identification of collaborations amongst actors. Clearly, this amounts to an indirect measure of actual networks of alliances and collaborations. It is clear that productive alliances, license agreements or even development collaborations are far from systematically leading to patenting. Moreover in the DBA database, the actual content of such collaborations are undoubtedly science and technology driven. Yet this does not prevent the analysis as we are interested in describing patterns of scientific and technical collaborations within the intra-paradigmatic construction of the biotechnology space. Co-patenting is all the more convenient and relevant in analyzing the co-evolution of knowledge and the network structure. Still, a further complication lies in the method of network description. Assessing networks is far from straightforward. For example, graph theories have proven extremely powerful in picturing networks of all sorts, including interactions among technologies or actors. Furthermore, they provide network measures of centrality and density (Burt, 1976; Freeman, C., 1991; Freeman, L., 1979) that are becoming gradually used in the economics of network literature (Orsenigo, *et al.*, 1998). But recalling our propositions, we shall essentially depict populations of co-patenting. Consequently, each patent has been described by the number of actors involved together with the description of their institutional type, namely public or private. Altogether, we have identified 7,205 co-patent applications, of which 3,985 have been characterized by both the numbers and the public or private type of partners involved.

Figure 1 describes the whole process of sample construction throughout propositions 1-4. From the original 96,284 patent applications (set Ω_1), the condition of relevant dates (set Ω_2) and actor identification (set Ω_3) led to 67,464 relevant patents (set Ω_4). Over the whole period, 3,985 co-patents were identified (set Ω_5), of which 3,822 belong to the 1975-1998 period (set Ω_6). Because of the very nature of propositions 1-4, no parametric statistical test seems to be eligible to test the first four propositions. Since we are essentially interested in the evolution of specified indicators, a longitudinal analysis of descriptive indicators is used. Table 1 both summarizes our method and reveals the expected outcome according the intra-paradigmatic phase under consideration. For convenience only, the results are presented in tables rather than in graphs.

Table 1. Method, Indicators and Expected Outcome Through Propositions 1-4. According to the Intra-Paradigmatic Phase 1 or 2

	Indicators	Variable	Phase 1	Phase 2	Table
P 1	Evolution of the # of patents	PAT	Increase	Decrease	Table 2
	Evolution of the # of new arrangements	INNO	Increase	Decrease	
	Evolution of the Herfindhal index	HERFPAT	Decrease	Increase	
P 2	Evolution of the # of actors	ACTOR	Increase	Decrease	Table 2
	Evolution of the Herfindhal index	HERFACT	Decrease	Increase	
P 3	Evolution of the # of patents from public R&D	PUBPAT	Increase	Decrease	Table 2
	Evolution of the public R&D share in patenting	SPUBPAT	Increase	Decrease	
P 4	Evolution of global co-patenting	COPAT	Increase	Decrease	Table 3
	Evolution of public-private co-patenting	HYBCO	Increase	Decrease	
	Evolution of public-public co-patenting	PUBCO	Increase	Decrease	
	Evolution of private-private co-patenting	PRIVCO	Increase	Decrease	

In proposition 1 which expect a substantial slow down in knowledge creation, two indicators are used, the number of patents by year of application (PAT) and the number of new technological arrangements, or combinations, by year (INNO). Both shall be distinguished as they provide different types of information. The first one deals with pure knowledge creation, while the second one discriminates the newness in knowledge creation. New patents may be applied for because either they consist of a radical breakthrough or because they substantially refine existing technological arrangements. Another important implication of the latter is that it allows us to measure some sort of concentration of patent applications along technological arrangements. The Herfindhal index, defined as the sum of the squared patent share for all technological arrangements, shall be used for this purpose (HERFPAT).[2] In proposition 2, we simply count the number of actors present in DBA for each year (ACTOR). Again, this rudimentary count measure is completed by the Herfindhal index, now defined as the sum of the squared patent share for all patenting organizations (HERACT). Proposition 3 concerns the role of academia in patent applications. Two closely related measures are used, the first dealing with the number of patents emanating from academia (PUBPAT), the second presenting the relative share of patent applications from public research organizations in the overall patenting activity (SPUBPAT). Finally in proposition 4, the description of network structure shall be depicted. Three types of networks are surveyed: collaborations amongst public research organizations (PUBCO), collaborations amongst private firms (PRIVCO), and hybrid networks involving both public and private organizations (HYBCO).

4. RESULTS

Table 2 reports the evolution of knowledge production and the actors involved in the biotech industry, associated with propositions 1, 2 and 3 respectively. Focusing on the evolution of the patenting activity between 1975 and 1998, we observe an increasing rate of patent applications. As could be expected, encouraging promises of biotechnology led to a corresponding exponential rise of patent applications (PAT). The emergence phase did not lead, however, to a slow down of patent applications. Rather, the rate at which patent applications increase remains continuously quite high. In other words, knowledge creation as expressed in patent applications does not express a yet expected declining phase. This crude observation should be taken with caution. The evolution of the total number of patents presents the rhythm of the global knowledge production regardless of the technological novelty held in the patent. The evolution of the variable INNO provides us with the information needed. It shows that biotechnological novelty followed a rapid growth before 1980. Reaching its highest mark in 1981, its level remains steady for the remaining period at stake. That is in level, the number of new technological arrangements average around 120 new combinations each year introduced in the technological landscape.

These observations lead us to confirm the non-decreasing rate of knowledge generation, thus invalidating our first proposition. However, there exist various reasons to remain cautious about this conclusion. Our first reason is purely technical and reflects a database artifact. DBA introduced two new technology classes in 1992, thus boosting the variety of potential arrangements within the dataset. Though we took care in building an a-temporal nomenclature, the effect of increasing variety remained. Ignoring this artifact would lead to the opposite conclusion. Second, the share of novelty (INNO) relative to patent applications (PAT) is constantly decreasing. This means that a greater part of knowledge generation is based on existing arrangements within the technological landscape. Third, the Herfindhal index (HERFPAT), which describes the concentration of technological arrangements, decreases until 1991 revealing the novelty of knowledge created. New technological arrangements are introduced in the technological system as scientific hypotheses are explored. However after 1991, the concentration of patents by arrangements is significantly increasing, i.e. HERFPAT rises. Consequently, the first phase corresponds to the expansion of the technological landscape by explorative learning strategies, while the second is characterized by the exploitation of existing technological arrangements. In the latter period, new technologies tend to only marginally improve the existing knowledge base of the industry within the more general framework of the established dominant design.[3] Altogether, though we remain silent regarding the rhythm of knowledge generation per se, we do see a more established landscape.

Table 2. Descriptive Statistics for Propositions 1-3.

	PROPOSITION 1			PROPOSITION 2		PROPOSITION 3	
	Reference Subset (Ω_2)			Reference Subset (Ω_4)		Reference Subset (Ω_4)	
Year	**PAT**	**INNO**	**HERFPAT**	**ACTOR**	**HERFACT**	**PUBPAT**	**SPUBPAT**
1975	27	7	68.88	15	108.03	9	0.47
1976	53	13	59.45	31	42.74	17	0.44
1977	106	24	35.19	49	25.62	35	0.52
1978	165	30	26.31	85	16.09	59	0.50
1979	475	62	21.43	146	13.84	128	0.44
1980	1192	105	18.40	264	8.66	297	0.39
1981	2558	168	15.29	381	9.24	519	0.29
1982	3066	134	16.07	443	8.17	650	0.29
1983	3683	130	16.19	528	6.72	802	0.29
1984	3862	132	17.76	563	7.21	833	0.27
1985	3829	73	19.38	583	7.32	845	0.28
1986	4043	100	19.43	641	5.61	871	0.26
1987	4655	77	20.72	715	5.15	1122	0.28
1988	4961	98	22.70	766	4.37	1354	0.32
1989	5028	97	24.45	803	3.87	1333	0.31
1990	4862	99	27.58	807	3.78	1187	0.29
1991	4730	138	16.96	778	4.01	1106	0.29
1992	5147	127	23.43	802	3.77	1249	0.30
1993	5167	127	27.78	821	3.81	1392	0.32
1994	5420	110	34.58	839	3.96	1680	0.37
1995	6322	117	36.53	852	4.16	1965	0.37
1996	6119	134	41.95	786	4.51	1912	0.39
1997	7476	142	41.42	761	5.69	2026	0.38
1998	7875	133	43.25	670	5.81	2002	0.40

Proposition 2 is more clearly corroborated. After an exponential rise in the number of actors involved in knowledge production in biotechnology until 1990, the total of actors stabilizes between 1989 and 1995 at around 800 and then decreases (PAT). Thus, a gross correlation is observed between on the one hand, the entrance of new actors with the emergence phase, that of radical advances in biotechnology, and on the other hand the exit of actors with the establishment of an identified dominant design. The latter operates as a turning point. While it greatly reduces both product and market uncertainties, it further designs research activities around clearly perceived technological complementarities. Furthermore, as innovation becomes closer to the market, firms tend to develop products and services through internal research to maximize the potential returns to investment. The Herfindhal index (HERFACT, thus concerning actors) reaches a minimum in 1992 to subsequently increase. Knowledge production in biotechnology is more and more concentrated around given combinations and major actors, though the technological landscape is still following an expanding path.

Following Mertonian principles as guidelines (David, Mowery and Steinmuller, 1994), proposition 3 states that as innovation gets closer to the market, academic organizations become less active in knowledge creation (PUBPAT, SPUBPAT). The proposition is not verified since the share of public research organizations in patent applications is continuously increasing from 1975 to 1998. At first glance, the observed advancement of the public R&D patenting is extremely puzzling. As pointed out by Dasgupta and David (1994) or Callon (1994), commercialization of science ought not to be the central mission of public research. Since both technological arrangements and the number of actors reach a steady state, we would then expect public research laboratories to withdraw from the scene. Moreover in the meantime, firms are expected to take the lead in biotechnology development as they are approaching the market, hence generating turnover and profit. In fact, our intuition is as follows: this trend illustrates the commercialization of science based on technological transfer from discovering scientists to those who will develop it commercially (Zucker, Darby and Armstrong, 2001). Here, the patenting of academically produced knowledge is one of the devices used by public policies to boost technology transfer and encourage firm creation by scientists. This phenomenon, which conforms to our own knowledge of the field, may be one of the explanations of the unexpected continuous growth of the share of patent applications by public institutions.

Table 3 presents the results on the evolving technological network structure (Proposition 4). We observe a positive trend for all actors until 1995. Since then, co-patenting declines rather sharply. Focusing amongst the institutional types of partners involved in collaborations (HYBCO, PUBCO, PRIVCO) reveals that this trends concerns all actors irrespective of the organizations involved in networking. This corroborates our proposition of a sequential increase and decrease in collaborations. A finer look at the figures leads to two additional observations. First, private technological alliances constitute the core of networking until the nineties (PRIVCO). As firms seek to appropriate knowledge and translate it into private rents, the role of R&D collaborations in innovative activities is diminishing since the beginning of the nineties. This goes together with the increasing share of individual patents (except in 1998). Second, collaborations amongst public research organizations become more important in the most recent period (PUBCO). We have previously mentioned the increasing role of public research in market-oriented knowledge generation activities. Its strategy becomes closer to that of firms in term of appropriation of scientific knowledge. We thus conclude that the emergence of public patenting has been accompanied by a significant increase in collaborations amongst public laboratories, though their relative share in global public patent application declines.

Table 3. Descriptive Statistics for Propositions 4.

	INDIVIDUAL PATENTS			CO-PATENTING			
	Reference Subset ($\Omega_4 - \Omega_6$)			Reference Subset Ω_6			
YEAR	PRIV	PUB	TOTAL	HYBCO	PRIVCO	PUBCO	TOTAL
1975	5	9	14	0	0	0	0
1976	20	15	35	0	1	1	2
1977	29	30	59	1	1	2	4
1978	56	48	104	1	0	5	6
1979	158	112	270	0	1	8	9
1980	444	271	715	6	6	10	22
1981	1220	470	1690	18	17	15	50
1982	1540	592	2132	8	14	25	47
1983	1931	749	2680	19	25	16	60
1984	2193	728	2921	30	29	36	95
1985	2099	750	2849	37	33	28	98
1986	2285	753	3038	28	64	44	136
1987	2596	940	3536	59	75	55	189
1988	2615	1129	3744	69	103	72	244
1989	2685	1110	3795	73	75	70	218
1990	2692	951	3643	71	79	79	229
1991	2518	896	3414	71	81	64	216
1992	2695	988	3683	81	83	86	250
1993	2615	1065	3680	110	93	101	304
1994	2640	1234	3874	116	75	149	340
1995	2993	1498	4491	135	102	150	387
1996	2728	1436	4164	119	71	168	358
1997	3014	1628	4642	112	61	132	305
1998	2802	1667	4469	85	50	118	253
TOTAL	44573	19069	63642	1249	1139	1434	3822

Our results could be summarized as follows. The evolution from emergence to maturity reveals the changing nature of technological exploration activities. Novelty is based on the creation of new arrangements reflecting the investigation of new scientific hypotheses. This phase was largely supported by the growth of scientific collaborations as revealed here. Note the substantial part of inter-institutional networks (HYBCO) in which firms and academics laboratories are involved. The transitional or maturing phase is based on arrangements within existing combinations i.e. existing but disjointed technological solutions within a given technological trajectories. In the meantime during the most recent years, networking activities seem to decrease. Thus, relying on extra-organizational partners in research activities tends to be phase-specific. That is, actors share complementary knowledge and competencies only when the technological uncertainty and the rate of novelty are high. When the technologies stabilize, they develop more traditional forms of organization in research activities.

Two final remarks are worth noting here. First, as biotechnology appears to be a vital competence for innovation in life sciences, patents play a central role in the firms' strategies. After the 1980's, the systems of intellectual property rights have gradually reckoned the economic potentialities of biotechnology. At the level of firms, stabilization and patenting have been included in firms strategies, especially in start-up strategies. Indeed, as pointed out by Afuah and Utterback (1997), patents are one of the central assets needed to succeed in the emerging (fluid) phase. Dodgson (1991) and Rabinow (1996) underline that one of the conditions sine qua non for the creation of start-ups is the opportunity to protect their innovation. Thus the observed evolution of patent and co-patent applications reflects more profound phenomena that go far beyond the fact of mere appropriation issues. It also expresses some form of the adaptive efforts of a wider set of actors, namely regulation institutions, governmental policy makers, the cultural adjustment of and in academia, not to mention other aspects regarding the demand side. At an aggregate level, technical change necessarily involves the development of new forms of organizations and institutions coordinating the local and bounded activities of a wide variety of actors.

Second, the observed declining amount of technological relationships does not necessarily deny the persistent role of collaborations. First DBA may principally reveal the first wave of biotechnology, centered on genetic sequencing and gene manipulation. And indeed, the observed evolution is well argued by the arrival of a set of stabilizing research procedures and instruments, such as PCR, bioinformatics, crystallography, etc. However, recent developments in new promising areas such as genomics, proteomics, pharmaco-genomics and/or single nucleotide polymorphisms are still in their infancy. We thus expect additional technological collaborations to come up in the coming years. Second recall that we are focusing on only technological collaborations expressed through co-patenting. In fact, there exists a wider range of reasons for which firms, and more globally organizations, should enter into alliances (Hagedoorn, 1990). For example, the pharmaceutical industry is still witnessing a substantial number of inter-organizational ties, involving a wide variety of actors: DBFs, large pharmaceutical firms, and public laboratories. Large firms rely heavily on university ties to improve their own knowledge base, while they call on DBFs to develop and explore new technical possibilities. Thus, our contention is that networks may reflect the changing nature of the technological landscape not in the mere number of ties, but rather in their types. In other words, networks are not dead. Their nature is simply changing together with the industry life cycle.

5. DISCUSSION AND CONCLUSION

This chapter has analyzed the evolution of network of technological collaborations during the life cycle of the industry. Based on the example of the biotech sector, it questions the network of collaborations as new forms of organizations. Powell *et al.* (1996) argue that in a field of rapid technological development, the locus of innovation is found within networks of interorganizational relationships that sustain a fluid and evolving community. Our empirical analysis refreshes Powell's picture of networks of collaboration, in that collaborative networks are the locus of innovation during the exploration phase, while in a more stable environment, actors tend to rely more on internal learning strategies. As revealed here, collaboration is indeed the industrial organization form, which allows the different actors to explore a large number of hypotheses and develop explorative learning in a highly uncertain environment. Such an organization reduces risk and cost, while enhancing the sharing of knowledge, competencies and scientific results. Thus, interdependencies promote a sense of community-level mutualism (Barnett and Carroll, 1987). As the technological conditions change and as the sector matures, technological trajectories appear and firms innovate within given trajectories. The degree of novelty is reduced compared to its former level. Firms develop incremental innovation combining existing technologies that they acquire through networking in a recent past.

However, during the exploitation phase, innovation is based more on internal R&D than in the unstable phase. On the one hand, firms already develop research capabilities and they are able to organize and perform independent research. On the other hand, as innovation is closer to the market, products and process directly compete within the market and co-operation is more difficult to develop. Networks still exist but the forms and the partnership within the network is changing. It is vital for firms to design collaborations with material and equipment providers as well as commercial networks. Whereas scientific and technological hypothesis have been explored through collaborations with diverse organizations, i.e. start-ups, large firm, dedicated biotech SMEs and academic labs, exploitation of few promising technological trajectories are developed through firms' internal capabilities. The initial scientific and technological initial accumulation creates barriers to entry because of irreversible investments and because of the cumulative nature of knowledge.

Our results are coherent with the ones obtained by Nesta and Saviotti (2002) when they underline the contrasted evolution of the breadth and the coherence of the knowledge base. They reveal two different patterns of knowledge creation and two opposite types of learning. In the fluid phase, it is difficult to predict the ways by which knowledge develops. Firms have to cope with high uncertainty and explore an expanding technological landscape by diversifying their knowledge base. Though in the short run, this reduces the overall coherence of the knowledge base, it mostly enhances higher adaptive capacity to several although poorly identified technological trajectories. As technology matures, firms are likely to abandon less promising competencies to concentrate on more fruitful competencies. Importantly the focus on fewer hypotheses that leads to the dominant design creates new barriers to entry for potential incumbents because of the shift from competence destroying process in the emerging (fluid) phase to creative accumulation in the maturing (transitional) phase.

In the meantime, the overall industrial configuration is changing: the market introduction of innovation leads to more rivalry among existing competitors. Winners win new markets and grow while the losers fail. Thus size becomes more and more important in reaching a critical minimum size to undertake costly and heavily instrumented R&D activities. The growth of firms results from internal innovations and from mergers and acquisitions, which participate to the reduction of the number of actors. As bio-industries become more mature, there exists evidence of a shift towards consolidation: 198 mergers and acquisitions have been reported in the past five years, 50% of which occurred in the last two years (Ernst&Young, 2001). The positive outcome of those mergers and acquisitions (M&A) is likely to be the emergence of an elite of European industrial biotech leaders on which the sector could grow and compete with that of the US, which is already mature.

Public policies need to adapt to the phase of the life cycle of the industry and to prepare the entry into the next step. During the 90's in Europe, public policy had mainly focused on three objectives: (1) the building of basic research capabilities to boost the rhythm of discovery in Europe; (2) commercialization of R&D essentially by supporting the creation of start-ups and (3) the building of common rules of the game to set up of a supporting environment with the harmonization of IPR legislation and the creation of stock exchange for high speed growing firms. And it is quite successful. In less than ten years, the number of European firms is comparable to that of US ones. However, a rapid glance at the US biotech firms is in this vein extremely informative. Whereas US and Europe have a similar number of firms, European ones remain significantly smaller and weaker. First, regarding their size, European biotech SMEs represent one third of their US counterparts (61,104 versus 162,000 employees). Second, fewer companies have gone public (105 vs. 300). Their market capitalization adds up to only one fifth of their US counterparts. Third, the R&D spending and revenue streams of European firms lag far behind those of the US.

APPENDIX 1. THE DBA DATABASE AND TECHNOLOGY CLASSES

The Derwent Biotechnology Abstracts (DBA) database defines biotechnology as: "The application of biological organisms and molecules to technical and industrial processes, i.e. the application of micro-organisms, animal or plant tissue cultures, enzymes, or any other cellular or sub-cellular biological system, to the production of pharmaceuticals, pesticides, chemicals, enzymes, single cell protein, energy, etc., on an industrial scale. Includes both de novo biosynthesis and specific biochemical modification of previously synthesized compounds." The following scientific items are covered in DBA: "Molecular biology and genetics, recombinant DNA techniques, cell fusion in plants, animals and microbes, animal and plant cell and tissue culture, micro-manipulation of embryos, molecular and cellular immunology, enzyme, organelle and whole cell immobilization, protein and enzyme engineering, fermentation technology and process engineering, including computerization." For each patent, the database indicates the patent holder (corporate affiliate), the year of applications and the technological areas to which the patent has been assigned. Because areas are listed, the conversion of the database into a workable format was needed. Notably, rather than having a list of the technological areas covered in a given patent, a vector of thirty technologies was assigned to all patents. This provided us with the raw material for further empirical manipulations. Descriptive statistics for each technology classes are found in table A1, covering the period 1981-1997.

Table A1Descriptive Statistics for DBA Technological Occurrences

Classes Principales	Classes Secondaires	*Mean Frequency*	*Global Growth Rate*
A-genetic-engineering-and-fermentation	*A1-Nucleic-Acid-Technology* *A2-Fermentation*	*60.5*	*+ 482.1*
B-engineering	*B1-Biochemical-Engineering*	*6.7*	*- 38.1*
C-analysis	*C1-Sensors-and-Analysis*	*3.2*	*+ 42.7*
D-pharmaceuticals	*D1-Antibiotics* *D2-Hormones* *D3-Peptides-and-Proteins* *D4-Vaccines* *D5-Other-Pharmaceuticals* *D6-Antibodies* *D7-Clinical-Genetic-Techniques*	*54.8*	*+ 555.0*
E-agriculture	*E1-Biological-Control-Agents* *E2-Plant-Genetic-Engineering* *E3-Pesticides* *E4-In-Vitro-Propagation* *E5-Agricultural,-Other*	*9.6*	*+ 669.6*
F-food	*F1-Food-and-Food-Additives*	*9.6*	*- 15.8*
G-fuels,-mining-and-metal-recovery	*G1-Biofuels-and-Solvents* *G2-Mining-and-Metal-Recovery*	*13.6*	*- 28.0*
H-other-chemicals	*H1-Polymers* *H2-Chiral-Compounds* *H3-Miscellaneous-Compounds* *H4-Polyunsaturates*	*6.4*	*+ 185.6*
J-cell-culture	*J1-Animal-Cell-Culture* *J2-Plant-Cell-Culture*	*16.0*	*+ 229.4*
K-biocatalysis	*K1-Isolation-and-Characterization* *K2-Application*	*25.6*	*+ 75.8*
L-purification	*L1-Downstream-Processing*	*7.5*	*- 77.0*
M-waste-disposal-and-the-environment	*M1-Industrial-Waste-Disposal* *M2-Environmental-Biotechnology*	*7.1*	*+ 103.0*

Source: Derwent Biotechnology Abstract.

Chapter Twelve
NETWORKS IN THE KNOWLEDGE ECONOMY RESTRUCTURING VALUE CHAINS:
Impacts of the Internet

Graham Vickery
Information, Computer and Communications Policy Division, Directorate for Science, Technology and Industry
OECD

1. INTRODUCTION

Information and communications technologies (ICTs) are becoming increasingly ubiquitous, and firms, industries and countries are reaping increasing benefits from their ongoing investments in ICTs and the more widespread use of the Internet. These benefits are continuing despite questions about the existence of a "new", knowledge-based economy, in light of the recent sharp price decline in technology stocks and the severe slump in the ICT equipment industry. The diffusion of ICTs throughout the economy has enhanced economic efficiency and the ICT-producing and ICT-using sectors have accounted for the bulk of overall productivity growth in a wide range of OECD countries (see OECD 2002a, 2001a and 2001b).

One of the principal new tools for enhanced efficiency and productivity growth at firm-level comes from the application of the Internet and associated electronic commerce strategies to many business processes within firms and between firms. These electronic networks are changing the nature of interactions within and among firms, and they have very large potential to transform the innovation process.

To gain a better understanding of some of the impacts of electronic networks and the Internet on business processes, this chapter explores the extent of the transformations experienced by established firms in established industries across the whole economy. Most of the firms in the sample are not directly connected with the supply of ICTs or the supply of Internet-related goods and services and were in existence before the Internet stock market bubble.

In contrast, much of the previous empirical work on the potential and actual impacts of the Internet has been focused much more narrowly. It has focused on companies closely linked to the Internet, for example those providing Internet communications equipment, Internet-related software, security or e-business services, Internet service providers and portals, or Internet content providers (see OECD 2002a).[1] Or else it has tracked the spectacular rise and fall of new Internet-based market entrants servicing retail and consumer markets (the "dotcoms") which were largely unsuccessful in testing a range of new business models built around the Internet as an ordering and delivery vehicle.

2. RATIONALE

The rationale underlying the empirical work described in this chapter is that:

- Widespread and continuing strategic and organizational shifts are taking place in the networks of business relationships. These wider shifts are in part driving Internet and electronic commerce strategies of firms, and are in part driven by Internet and electronic commerce strategies of firms. Changes are taking place all along the various stages of value chains from product innovation (product design, development, differentiation, and customization) all the way through to market search and development, marketing, logistics and customer support.

- The Internet and associated electronic commerce strategies are enabling firms to:

1) Change the nature and intensity of their existing relations with suppliers and customers for all kinds of business inputs and outputs in business areas of research, design, development, production and marketing along specific sectoral value chains; and

2) Develop new relations with suppliers and customers for inputs and outputs.

- Many of the likely directions for the development of electronic commerce and Internet-related strategies can be assessed by studying pro-active firms and early adopters among established firms in a period when electronic commerce and Internet strategies are in their early and innovative stages.

- Study of the uptake and use of the Internet and e-commerce by early adopters is likely to give a consistent view of medium-term impacts of business Internet and electronic commerce strategies as they begin to diffuse more widely among businesses in most industries and size classes.

- The transformation of the business sector and impacts on businesses of the Internet and electronic commerce are still expected to be large, despite the collapse of the "dotcom" and telecommunications investment bubbles (OECD 2002a).

3. OBJECTIVES AND METHOD

The results described in this chapter draw on part of the work prepared for the OECD electronic commerce Business Impacts Project (EBIP). This was designed to:

- Explore the rapidly evolving adoption and use of e-commerce and Internet strategies
- Look at the implications of e-commerce for firm strategies
- Examine impacts of e-commerce and the Internet on sectoral dynamics and market structure
- Draw out e-commerce and Internet-related policy issues

The overall objectives of this work were to improve understanding of the impacts of the early adoption of Internet strategies and electronic commerce on business, using a methodology based on face-to-face firm interviews. Most detailed firm-level empirical research carried out previously has been anecdotal, fragmented and not comparable across sectors or countries. This is despite awareness that the major impacts of the Internet are on established firms in established industries, and increasing policy interest in these wide-ranging and pervasive impacts on the productive economy. By developing, testing and applying a common methodology for firm-level case studies, the work aimed to improve the comparability of studies across firms, sectors and countries.[2]

To provide insights into likely developments across the whole economy, the project focused on e-commerce and Internet applications by established firms in established industries operating mainly in business-to-business electronic commerce that were proactive early adopters of e-commerce strategies. Start-ups and consumer-orientated firms were not the focus of the study. A further objective was to capture as many changes as possible in the networks of business relations conducted over electronic networks. To this end, a wide definition of electronic commerce was adopted to include all electronic transactions conducted over computer-mediated networks.[3] An important part of the work was designed to assess migration from legacy e-commerce systems such as Electronic Data Interchange (EDI) to Internet-based systems.

The work was designed to exploit the common methodology for firm-level case studies to build a detailed database of the sample of firms to improve comparative analysis of these firms. This database was constructed to include information on firm characteristics, adoption and impacts of the Internet and electronic commerce strategies across different sectors to improve understanding of these pro-active adopting firms. These data could then be used to achieve a further set of analytical objectives including:

- Preparing a set of sectoral, national and cross-country research reports based on a common analytical framework,[4]
- Improving understanding of firms' views of what are important government policy issues related to electronic commerce, and
- Drawing policy implications applicable to the wider context for electronic commerce and business applications of the Internet.[5]

4. CONCEPTUAL FRAMEWORK

Electronic commerce is defined widely to provide a broad overview of how electronic networks are affecting innovation processes in firms.

- The development and application of ICTs has made it possible to conduct complex commercial transactions electronically over communications networks. Electronic *transactions* over computer-mediated networks are defined as electronic commerce. Transactions over the Internet do not make up all of electronic commerce, but this is the key technology in its recent evolution. Electronic commerce also encompasses other well established but more closed computer-mediated systems, most notably Electronic Data Interchange (EDI), which had been carried out over proprietary systems linking a limited and carefully restricted group of transactors.

- The overall framework for the empirical research and analysis is based on a structured set of transactions. Transactions are defined widely as any exchange between participants in a market. In this analysis transactions are classified into three broad groups, involving activities preceding transactions, the actual transactions and activities supporting transactions:
    - *Transaction preparation* (advertising, catalogues and stock lists, information services, negotiation);

- *Transaction completion* (ordering, billing and payment, finance, delivery);
- *Transaction support* (information capture, information management, market analysis, market development).

The development, adoption and implementation of electronic commerce and Internet strategies are also innovations in how transaction preparation, transaction completion and transaction support activities of firms are organized. Drawing from the extensive literature on firm-level innovation, the activities related to the transactions of firms can be cross-classified into three different kinds of innovations:

- *Product innovation* (diversification, differentiation, customization, bundling of products and services);
- *Process innovation* (design integration, logistics, production, co-ordination of administration and production processes);
- *Relational and organizational innovation* (geographical expansion, market segmentation, establishing trust, increasing loyalty).

The results of the firm-level interviews are described in the following parts of this chapter in terms of firm-level transaction structures and firm-level innovations. The adoption of Internet-based and e-commerce strategies results in major changes in the way that firms organize their businesses and business processes, and this is reflected in changing transaction structures and innovation processes.

5. RESULTS

This work draws on a database of 217 reporting firms, and 30 sector reports across 14 broad sectors in both physical and intangible products in 11 participating countries.[6]

The empirical studies were designed to cover both tangible and intangible-producing sectors, in participating OECD countries over a short time frame. Although the focus was primarily on sectoral value-chains, the study was designed to have a reasonable distribution of firm sizes to enable the importance of firm size to be assessed. Furthermore by including a range of countries (large G7 countries, smaller Northern European countries, countries still catching up economically) some aspects of the extent to which country-specific characteristics are important could be assessed, particularly in policy areas. The studies were mainly undertaken during the second half of 2000 and early 2001. Design and testing of the methodology were launched in late 1999 with ten countries participating: Canada, France, Italy, Korea,

Mexico, the Netherlands, Norway, Spain, Sweden and the United Kingdom. Portugal joined subsequently.

The sectoral distribution of the firms studied was approximately: firms/sectors producing physical-tangible products and/or with physical-tangible assets were 70% of reporting firms; firms/sectors producing services-intangibles products and/or with intangibles assets were 30% of reporting firms. (See Table 1 for the sectoral distribution of sector value-chains studied cross-classified by ISIC sector). The sectors studied producing tangible products were textiles and garments, book selling, chemicals and plastics, glass, steel, non-ferrous metals, ICT-goods, automobiles, and construction. The sectors studied producing intangible products were travel and tourism, transport and logistics, banking and financial services, business services, and music. The size distribution of the sample consisted of 59% of the sample of 217 firms had more than 250 employees, while 41% had less than 250 employees.

6. WHAT IS MORE IMPORTANT IN THE SUCCESSFUL DEPLOYMENT OF E-COMMERCE AND THE INTERNET, THE 'COMMERCE' FACTOR OR THE 'E' FACTOR?

E-commerce and Internet strategies are part of much larger business and economic transformations. The empirical and analytical work undertaken suggested that successful e-commerce applications and Internet use by firms are usually part of much broader strategies to respond to business challenges. E-commerce business strategies are led more by commercial than technological considerations, and successful e-commerce strategies are led by commercial considerations. The 'commerce' factor outweighs the 'e' factor.

Nevertheless, e-commerce represents a major innovation in doing business that all firms will have to adopt. Successful firms are the ones with a coherent overall business strategies with a major emphasis on both e-commerce and IT skills development embedded in these strategies.

7. WHY ADOPT?

Motivations to apply e-commerce and Internet strategies are high where ICT investment is already large and risk low. Most firms in this study had business objectives related to their e-commerce and Internet strategies such as:

- Reduce costs
- Increase transaction speed and reliability

- Improve management capabilities
- Develop and/or improve collaboration capabilities
- Create interdependencies
- Manage customer relations better
- Create more added-value

8. WHAT IS GOING ON-LINE?

Firms are cautious about putting important transactions and network relations on the Web and conducting transactions over the Internet. When they do so they protect their transactions and advantages.

Differences in technology use oriented to customers or suppliers are slight. Most of the more significant variations are consistent with the degree of diversity that is to be expected in a sample of the kind described here. The largest variations concern Internet EDI, which is used primarily with suppliers, and Extranets, which are used primarily with customers. The general profiles of EDI and Internet EDI are nonetheless similar, and probably attributable to the generally proactive stances in the group of respondents. Likewise the Extranet and WWW profiles are similar. See Figure 1. This trend is confirmed in the reported e-commerce activities applied to both customers and suppliers. Figure 2 shows that the top four categories are mainly transaction preparation functions and the bottom four are mainly transaction completion functions.

The WWW and e-mail applications have become very significant in the technology portfolio of firms in commercial processes. The technology profile of most e-commerce implementers has widened to include most available e commerce platforms. However the *new* technology base is being directed primarily at transaction preparation functions, information services, catalogues, advertising. The limited role of traditional forms of EDI in transaction preparation would appear to have been overtaken by Internet EDI and Extranet applications. "Closed" applications in transaction preparation are now roughly equal to the number of "open" applications. The primary role of more open e-commerce architectures is to reach out to new suppliers and customers, or to provide general information about the firm and its products.

In transaction completion the technology configuration is biased decisively towards closed systems. Most EDI and Internet EDI activity is concentrated in the completion functions, historically the primary orientation of EDI. However, many firms reported that a commercial exchange, even between accustomed trading partners, could encounter a range of technologies at various times as new tools are added to old ones. This provides continuity and permits continued maximization of the installed base of technology, but it can also create legacy problems, many of which become acute where additions and/or subtractions occur in the customer or supplier base.

Overall, for successful implementation in the sample of pro-active firms studied:

- Transaction preparation (advertising, catalogues, information services) was more likely to use the WWW with customers. The Web is seen as an effective way of spreading information about firms and their products both intra-sectorally and inter-sectorally and geographically.

- Transaction completion (ordering, billing and payment, finance) is less likely to be on the Internet and WWW and much more likely to be on established EDI systems or EDI over Internet, or Extranets. (closed systems using TCP/IP protocols) with suppliers which are providing important product-related inputs (transaction completion).

- Migration to Web-based systems is partial and firms have been slow to move from established closed systems to Web-based and more open systems. Migration from EDI to Web-based systems, for example in ordering, is only partial. Many firms remain slow to move from their established closed and proprietary systems, which are secure and in which they have invested heavily towards more "open" systems, which may be more flexible and less expensive but which may also be more difficult to secure and more open to misuse. See Figure 3.

Figure 1. E-commerce Technologies with Customers and Suppliers

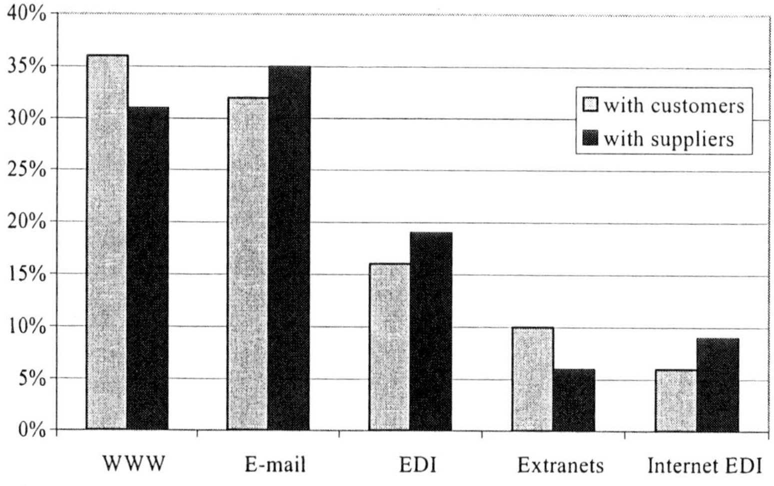

Source: Electronic commerce business impacts project, OECD, 2002.

Figure 2. On-line Activities by Transaction Type

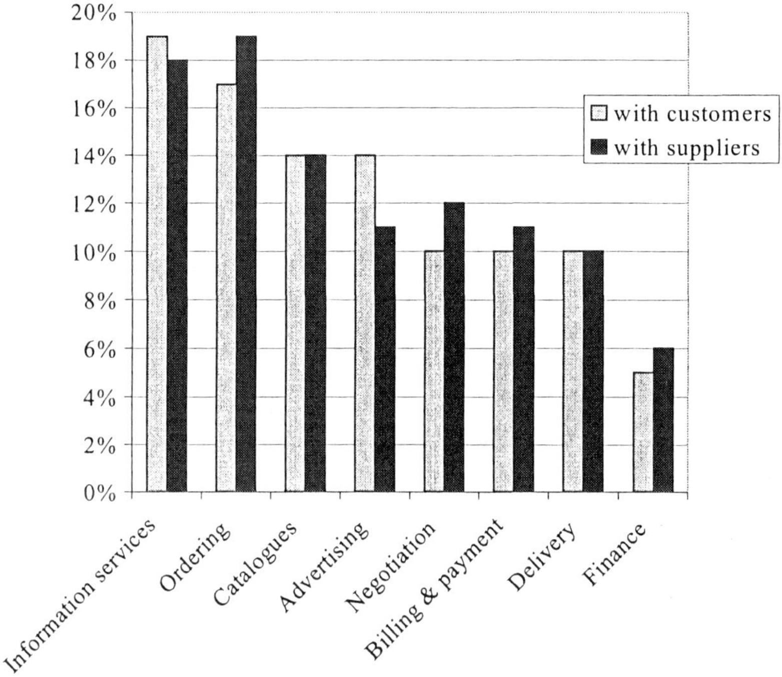

Source: Electronic commerce business impacts project, OECD, 2002.

Figure 3. The Use of Different Technologies for Different Transactions

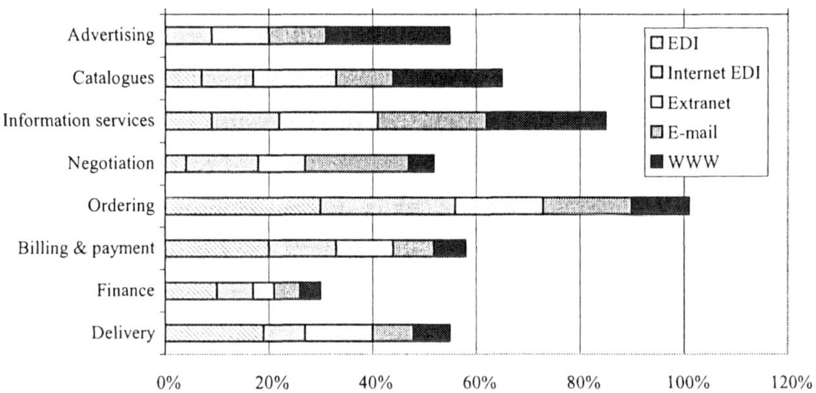

Source: Electronic commerce business impacts project, OECD, 2002.

9. WHAT ARE EXPECTED AND ACTUAL IMPACTS ON FIRM TRANSACTIONS?

On of the most critical and unknown areas in the whole process of e-commerce and Internet adoption is an assessment of the impacts of this adoption on firms. This section looks at expected impacts – where firms have been in most cases been optimistic about the number of areas and extent of impacts, and actual impacts as assessed by firms which have been in most cases considerably more modest.

Expected impacts: A very large majority of firms in the study reported that they expected e-commerce would have major effects on their business functions. These expected effects were clustered particularly in the area of transaction preparation (see Figure 4), where they were spread fairly evenly across different transaction preparation areas (information services, catalogues, advertising), as firms use e-commerce to make themselves better known and to organize their interactions with other firms. In addition, three quarters of firms expected effects in transaction completion (most notably in the area of ordering, where 65% of firms expected effects, the highest share), and two-thirds in production support (information functions such as information capture and management were the most important) (see Figure 4).

By *broad sector*, transaction preparation was the area with the highest expected effects for firms with tangible products and assets and firms with intangible products and assets. More firms with tangible products/assets expected effects in transaction preparation than in the whole sample (82% compared with 80% in the whole sample), most notably higher than the total sample in the area of catalogues and stock lists as well as in the transaction completion area of ordering. On the other hand, more firms with intangible products/assets were expecting effects in production support (72% compared with 68% in the whole sample). These were most notably higher than in the total sample in the areas of information management and market development, as well as in on-line product delivery.

By *firm size*, transaction completion was the most important e-commerce-related innovation for firms with more than 250 employees (78% compared with 73% in the whole sample of firms). This greater involvement in transaction completion by large firms was also across all categories of business functions but particularly in the area of ordering (71% compared with 65% in the whole sample). This suggests that large firms are more likely to be undertaking more widespread use of e-commerce, notably in transactions, and expecting effects there. In contrast, the largest proportions of smaller firms (equal or less than 250 employees) expected effects in the area of transaction preparation (87% compared with 80% in the whole sample) and production support (72% compared with 68% in the whole sample). The most notable differences from the whole sample were in the areas of advertising (71% versus 47%) and market development (50% compared with 37%), as small firms use e-commerce to make themselves known and explore new markets to a greater extent than large firms.

In most countries the highest proportion of firms expected effects in the area of transaction preparation, with well over two-thirds of firms expecting effects most notably in information services. Exceptions were Canada and Spain (transaction completion) and France (production support) with these national differences most likely due to the sectors studied.

Actual impacts: Actual impacts are at a lower level and follow the same pattern as expected effects. Around 60% of firms reported actual impacts, compared with around three-quarters of firms expecting effects. This lower level of actual impacts was true at aggregate level of transaction preparation, transaction completion and production support, and at all of the detailed levels of business functions (see Figure 4). The order of what firms described as the most important actual impacts was also very similar to the order of expected effects. The two most important impacts were in ordering (49% of firms in the database reporting impacts) and information services (47%) which were also the most important areas for expected effects.

There was relatively little variation by *broad sector* in terms of actual impacts on business functions. For firms with tangible products and assets, transaction preparation was the area with the highest number of firms reporting actual impacts (69% of the sample compared with 66% in the total). For firms with intangible products and assets, there were no great variations from the patterns reported in the total sample. For firms with tangible products there was little variation from the total sample in terms of detailed business functions, except they were generally higher than for firms with intangible products. Firms with intangible products generally were lower than for the total sample, except there were considerably higher impacts in areas where they could be expected -- information management, market development and on-line delivery.

For actual impacts by *firm size*, transaction preparation was the most important area for e-commerce-related impacts for firms with more than 250 employees. Large firm patterns were generally lower than for the whole sample. In contrast, smaller firms (equal or less than 250 employees) experienced higher actual impacts in the area of transaction preparation (76% compared with 66% in the whole sample), notably in advertising (63% compared with 40%) and catalogues (54% compared with 41%). Smaller firms in the sample appeared more likely to use e-commerce and the Internet to make them known. A relatively larger share of small firms also saw impacts in the area of production support (64% compared with 54%), in areas of information capture, market development and market analysis, as small firms use electronic networks to explore and develop markets.

In most *countries*, transaction preparation was the most common area of actual impacts, with more than one-half of firms experiencing impacts as for the whole sample. The exceptions were firms in the Netherlands and Spain (transaction completion) and France (production support), which behaved similarly to the pattern of expected effects.

Figure 4. Expected/Actual E-Commerce Impacts by Transaction

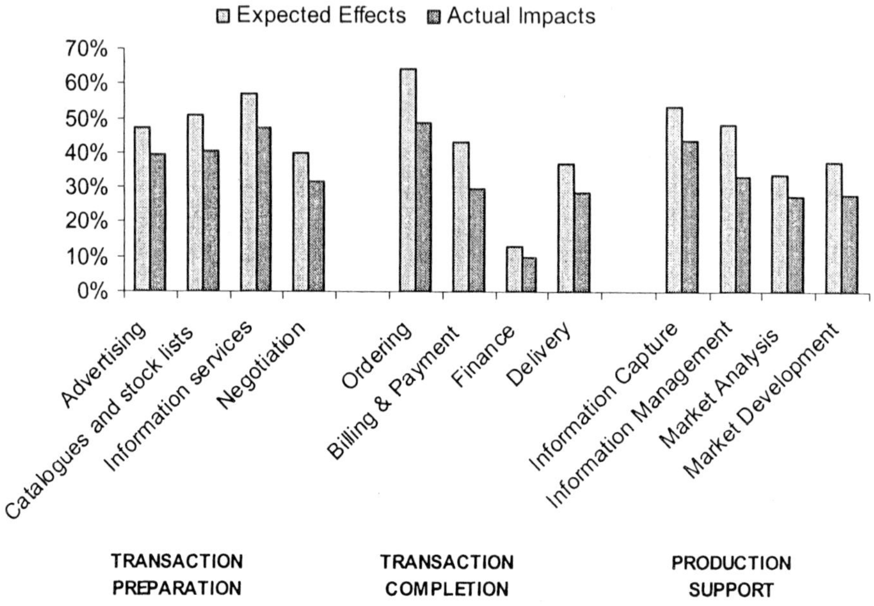

Source: Electronic commerce business impacts project, OECD, 2002.

10. WHAT ARE EXPECTED AND ACTUAL IMPACTS ON E-COMMERCE AND INTERNET-RELATED BUSINESS INNOVATIONS?

Firms were also asked what they expected to see and what they actually experienced in terms of product, process and relational (organizational) innovations.

Expected impacts: The majority of firms studied reported that they expected e-commerce would have major effects on their operations. These expected effects were clustered particularly in the area of process innovation and within process innovation in areas of logistics and co-ordination. Relational innovation was the next highest area where effects were expected by interviewed firms, particularly in the areas of geographical expansion and in building trust (increasing confidence in inter-firm relations) and loyalty (increasing the intensity and duration of inter-firm relations) (see Figure 5).

By *broad sector* process innovation was the area with the highest expected effects both for firms with tangible products and assets and firms with intangible products and assets. However, a higher proportion of firms with intangible products/assets expected effects in all three areas of product innovation (73% compared with 55% in the whole sample), process innovation (77% compared with 70%) and relational innovation (73% compared with 63%). This was particularly noticeable in the areas of product differentiation and diversification, process co-ordination, and geographical expansion, where a considerably higher share of firms producing intangible products expected effects from e-commerce.

By *firm size*, process innovation and the search for production efficiencies was the most important e-commerce-related innovation expected by firms with more than 250 employees (73% compared with 70% in the whole sample of firms). This was also in the same areas as for the whole sample, in logistics (51% compared with 47% in the whole sample) and co-ordination (50% compared with 42% in the whole sample). In contrast, the largest proportion of smaller firms (equal or less than 250 employees) expected effects in the area of relational innovation (69% compared with 63% in the whole sample). This was most notable in the area of geographical expansion (59% compared with 41%), as they expected to use electronic commerce to make themselves more widely known.

In most *countries*, the highest proportion of firms expected effects in process innovation, with well over one-half of firms expecting effects on processes. Exceptions were Italy (product innovation) and Portugal (relational innovation), but these national differences are most likely to be due to the sectors studied in both countries (notably banking in Italy, and glass production in Portugal), rather than unique national characteristics.

Actual impacts: Actual impacts are at a lower level and follow the same pattern as expected effects. A little less than one-half of firms overall reported actual impacts, compared with over 60% of firms expecting effects (see Figure 5). This lower level of actual impacts was true at aggregate levels of product, process and relational innovations, and at more detailed levels of innovation (see Figure 5). There were some changes in the order of what were described by firms as the most important actual impacts. But the two most important impacts were in co-ordination (34% of firms in the data base reporting actual impacts) and logistics (32%), with only their ranking switching (logistics was the most important expected effect, co-ordination second). Process innovations in the area of design were the area where there was the greatest similarities between actual impacts reported by firms in the data base and the prior effects that they expected.

By *broad sector* breaking impacts down by sector and firm characteristics, for tangible products and assets, process innovation was the area with the highest number of firms reporting actual impacts (56% of the sample). For intangible products and assets, product innovation had the largest number of firms reporting actual impacts (50%), and this was considerably higher than for tangibles (35%). However, more firms producing tangibles reported actual impacts in both process (56% against 48%) and relational areas (48% against 41%). For tangible products the process areas of logistics and co-ordination both had the highest actual impacts (35% of the sample), followed by geographical expansion (32%), whereas for intangible products product diversification was the most common (36%). In all cases the number of firms which did not or could not report detailed areas of actual impacts was considerably higher than those that did. For sectors where there are relatively large data sets, automobiles (tangible products) had a high level of impacts in processes (55%), particularly in co-ordination and logistics, and tourism and travel (intangible products) had both product and process impacts in product diversification and process co-ordination respectively.

By *firm size*, process innovation and the search for production efficiencies was the most important e-commerce-related innovation for firms with more than 250 employees (54%, the same as for all firms). This was in the same detailed areas as for the whole sample -- logistics (36% compared with 32% in the whole sample) and co-ordination (40% compared with 34%). In contrast, the largest proportion of smaller firms (equal or less than 250 employees) experienced actual impacts in the area of relational innovation (54% compared with 46% in the whole sample), most notably in the area of geographical expansion (44% compared with 30%). Product innovation impacts were also relatively important for smaller firms (44% compared with 39% in the sample). Overall, small firms were more likely to use e-commerce as part of the development of new products and to explore new markets.

In most *countries*, process innovation was the most common area of actual impacts, with around one-half of firms experiencing impacts in this area. The exceptions were Italy and Portugal (relational innovation was the most common area of actual impacts), but these national differences are most likely to be due to the sectors studied in both countries (banking in Italy, and glass production in Portugal), rather than to national characteristics.

Overall there are considerable differences by sector and perhaps by firm size although this may be linked more to differences in sector distribution and firm activities than being specifically related to size. Product innovations are more common among firms with intangible products and assets (mainly services firms); process innovations are more frequently implemented by large firms with tangible products. Market expansion and segmentation are more common for firms with intangible products and assets, and small firms particularly in intangibles areas benefit from Internet-based expansion strategies as they use e-commerce and Internet-based expansion strategies to make themselves known and explore new markets to a greater extent than large firms.

Figure 5. Expected/actual e-commerce and Internet-related business innovations

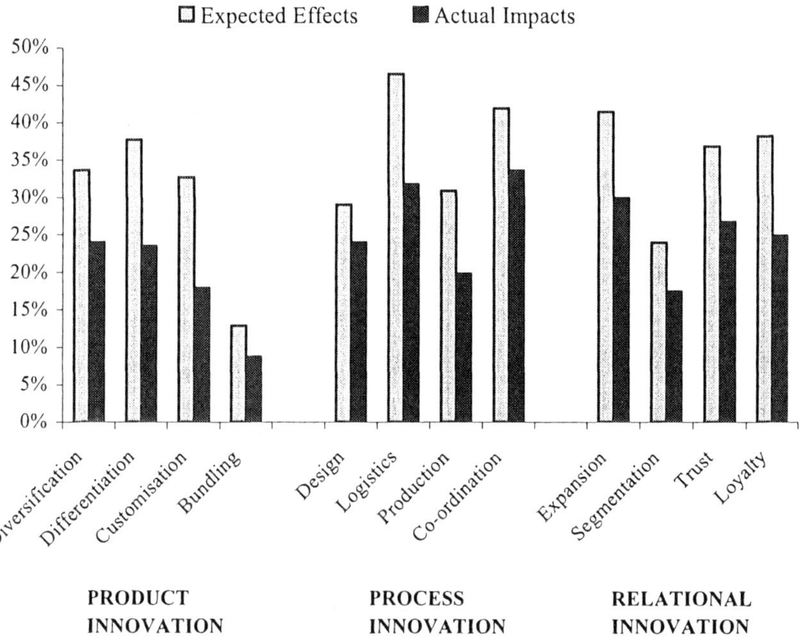

Source: Electronic commerce business impacts project, OECD, 2002.

11. MEASURABLE IMPACTS OF E-COMMERCE AND INTERNET STRATEGIES ON TURNOVER, PROFITABILITY AND EMPLOYMENT

A small but significant share of the firms studied reported quantitative impacts of e-commerce. Almost one third of the 217 firms studied reported positive measurable impacts on turnover and on profitability. None of the firms in the sample reported decreases, and around one seventh reported no change. However, well over one-half of firms were not able to report on impacts or were of the opinion that the question was not applicable, suggesting that it is too soon to measure impacts or that firms are not able to separately identify them (see Figure 6). Within the sample reporting impacts firms reporting positive impacts on turnover and those reporting on profitability were often different, with only about 16% reporting increases in both.

On the other hand, the impacts on *employment* were more evenly balanced, with the number of firms reporting positive and negative impacts both being low (11% and 8% respectively) but with more firms reporting positive impacts. About one-third reported no change and again the majority of firms were not able to report (see Figure 6). However many firms in the sample reported upskilling and changes in the composition of the workforce.

By *broad sector*, firms producing tangible products more commonly reported increases in turnover and profitability than in the whole sample (33% and 35% respectively), but more commonly reported decreases than increases in employment, with a large share of firms reporting no change. On the other hand, a smaller number of firms producing intangibles reported positive impacts on turnover and profitability (22% and 14%), but a much higher share reporting positive impacts on employment (17% versus 3% reporting decreases). This probably reflects the new opportunities that are presented for intangibles products to be organized and delivered electronically (financial services, business services, music) which leads them to expand operations but which is slower to increase cash flow and profits. This same pattern is seen in industries such as automobiles (over one-third of firms reporting increases in turnover and profitability) compared with travel (very few firms reporting either).

By *firm size*, large firms in the sample did not deviate significantly from the overall in reporting positive impacts except that they less commonly reported increases in turnover (22% compared with 29% in the whole sample). Small firms on the other hand more commonly reported increases in turnover (40% compared with 29% in the whole sample). By *country* the sample reporting positive impacts are too small to draw conclusions on significant differences across countries. However there were some national variations, with for example French and Canadian firms more likely to report positive impacts on profitability, and UK firms to report positive impacts on turnover and employment. This suggests that both sectoral composition of the national studies and business cycle effects may be at play.

Figure 6. Impacts on Turnover, Profitability and Employment

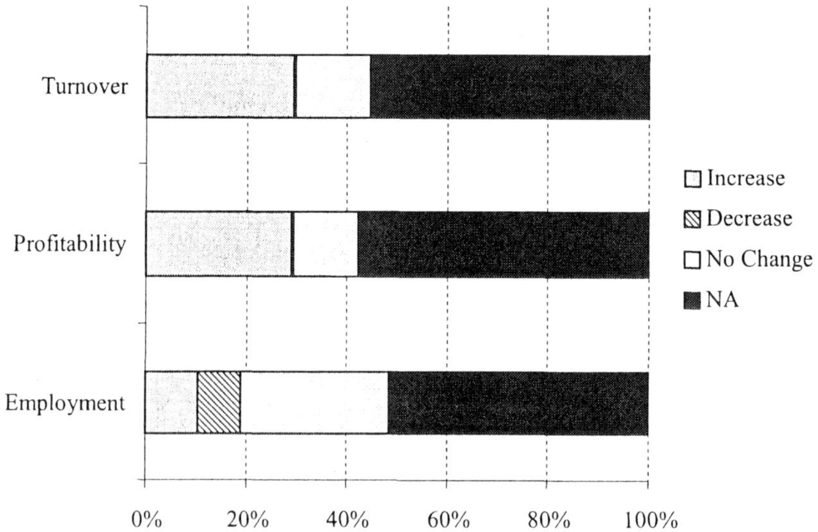

Source: Electronic commerce business impacts project, OECD, 2002.

A large proportion of the firms studied were experiencing general impacts in the value chain and broadly in terms of market organization (for example, market entry and exit, supplier and customer base). Over one-half were either experiencing impacts (49%) or thought that they would but that it was still too early (8%), whereas only one-fifth experienced no impacts and less than one-quarter (23%) did not see these impacts as applicable. However the kinds of impacts experienced in value-chains and market structure varied very widely and need further classification and analysis.

Overall, almost one-third of firms reported positive measurable impacts on turnover and on profitability, and none reported decreases. However, well over one-half of firms were not able to report on impacts or said the question was not applicable. The share of firms reporting employment impacts was lower, and more firms reported positive than negative employment impacts. However many firms reported upskilling and changes in the composition of the workforce. A crucial emerging issue is the organization of firms to use their human resources and other assets more effectively to achieve business goals.

12. MEASUREMENT ISSUES RELATED TO IMPACTS

The work mainly focuses on established firms and sectors, studying transformations within these firms and along the value-chains of these industries, concentrating on what are usually described as business-to-business transactions. These transactions and firms have been much less affected by recent events in the technology and Internet firm sector, and the bursting of the dot.com bubble. Furthermore, firms selected for interview were mainly pro-active firms and their suppliers and customers. It was believed that the changes taking place in these firms are very likely to be a leading indicator of the broader transformations likely to place in the sectors studied, and to a certain extent across the whole business sector.

Results of the firm studies showed that:

- Many firms in the study responded that it was too early for impacts to be assessed, as most firms were only now implementing e-commerce and Internet strategies, and they expected impacts to accrue in the future.

- Furthermore, many firms saw the impacts in a holistic way. They embedded Internet adoption and use as essential parts of wider firm strategies to improve efficiency and competitiveness. It is thus difficult to: i) disentangle impacts specific to firm-level e-commerce strategies from impacts of other changes in the firm and sector; ii) determine causality, for example successful firms may be more likely to adopt successful Internet strategies rather than successful Internet strategies driving overall firm success.

- Finally, the timing of this work must be kept in mind. The firm interviews were undertaken mostly end-2000 and early-2001, before the sharp downturn in OECD economies, particularly in the ICT sector. The downturn is likely to affect the kinds of impacts that firms are looking for from Internet and e-commerce strategies. These strategies are now more likely to be turning to survival, cost-cutting and boosting labor productivity, rather than for example product diversification and geographical expansion.

Overall the implications are that e-commerce adoption by pro-active firms has positive quantifiable impacts on turnover and profitability for a significant number of firms and more mixed impacts on employment, but that the majority of firms could not report impacts at all. The results also suggest that looking at Internet transactions only (buying and selling) masks much of the wider use of e-commerce and the Internet for a wide range of business functions.

13. WHAT POLICY ISSUES ARE IMPORTANT?

Firms identified competence factors (general education, specific IT/e-business skills) as being crucial to their involvement in e-commerce activities. Other factors consistently mentioned were costs including, but not limited to, technology. According to firms, in the area of confidence and trust, problems relate mostly to clarification, enforcement and cross-border inter-operability of existing legal frameworks, rather than a need to create new ones.

Overall, several broad areas were identified where the scope for government policy action is the greatest:

- Skills and competences
- Infrastructure (pricing, broadband)
- Market structure/competition areas

There is also continuing concern for small firm issues.

Table 1. Cross-tabulation between sectors studied and ISIC Rev.3 divisions

Sectors studied	ISIC Tabulation and Division
	A-B. Agriculture, fishing
	C. Mining
Textiles and garments	D. Manufacturing
	17-19 Textiles clothing, footwear
Publishing (music)	22 Publishing
Chemicals and plastics	24, 25 Chemicals, rubber and plastics
Glass	26 Non-metallic mineral products
Steel, non-ferrous metals	27 Basic metals
ICT-goods	29, 30 Machinery and equipment, computing machinery
Automobiles	34 Motor vehicles
	E. Electricity, gas, water
Construction	F. Construction
(Bookselling)	G. Wholesale and retail trade
Travel and tourism	H. Hotels and restaurants
Travel and tourism, transport and logistics	I. Transport and communications
Banking and financial services	J. Financial intermediation
Business services (fashion design)	K. Real estate, renting and business activities
	L-Q. Public administration, education, health, etc.

Note: Bookselling and fashion design are classified as sectors focused on tangible products and assets, music is classified as a sector focused on intangible products and assets.

Chapter Thirteen
TOWARD THE CAPTURE OF INNOVATION POTENTIALITY IN SOCIAL ENVIRONMENTS

Susan A. McDaniel,[1] FRSC
Department of Sociology
University of Alberta

There is something of what Pitrim Sorokin, the esteemed Russian émigré social theorist, has called the "Columbus complex" in many contemporary discussions of innovation. It is the rediscovery of something well known, only new to those who have never before seen it. That this may apply to innovation gives potential power as well as a delusionary cast to any ostensibly new analytical examination of innovation.

This chapter launches from two quasi-distinct literatures: one on innovation impacts in society, or more precisely, on the presumed beneficence of science and technology through innovation to economies and societies. The second literature is on social cohesion/ social capital/ social indicators in the "knowledge age" (see Stanley, 2001 for but one example). It is with a foot in each of these streams that much current policy thinking about innovation in Canada can be situated. The first literature takes a "hell bent" for innovation approach, almost desperately hopeful that beneficence, both economic and social, will tumble forth, if only we can become sufficiently new and different. That this is more a leap of faith than an empirical claim goes largely unsaid (McDaniel, 2000; Meisel, 1998; Wajcman, 2002).

The social cohesion/social capital literature concerns itself largely, thus far, with the impacts of global changes, including technological changes and innovations, on society: "...[T]here is a sense that the institutions and attitudes that have served as catalysts for cohesion, particularly at the level of the nation-state, are becoming less relevant as Canadian communities face the repercussions of globalization..." (Stanley, 2001, p. 4). Chief amongst these is the advent of a knowledge society that, purportedly is driven by technological change and innovation (see Castells, 1999). However, it is more of an assumption than an empirical verity that technological change will have social impacts. Indeed, some empirical and theoretical analysts have found that the causal direction might be the other way around, i.e. that it is social and economic change that creates demands for technological change and innovation (Wajcman, 2002; Stehr, 1994).

Both literatures as well as most popular accounts are imbued with technological determinism, "...with its hopes for a bright future promised as an evolutionary advance" (Lacroix and Tremblay, 1997, p. 4). This hope is, however, paralleled by "anxiety provoked by the invasion of these new technologies..." (Lacroix and Tremblay, 1997, p. 4). Fetishism emerges which reifies technology and innovation as change agents, but sees them as forces like gravity by which we are impacted, largely for good, but with some bothersome negatives. Technology is seldom conceptualized, at least in the policy literatures or in the growing research on innovation, as a social relation, a social product textured by the conditions of its creation in society and its social uses.

The assessment and capture of social impacts, however, on which there has been significant if beginning progress, can assist in the challenge of conceptualizing and ultimately capturing in statistical systems something about innovation potentiality in social environments. The question of what constitutes an environment conducive to science and innovation was a recurrent theme in Canada during the Science and Technology Review (June 1994 - March 1996) (see McDaniel, 2000; Meisel, 1998). That a complex, dialectical relationship exists between scientific advances and social changes has long been known, although perhaps forgotten or overlooked in much contemporary discussion. Technological deterministic thinking so much dominates that it has come to be accepted that even social problems may be solved by technological "fixes." The image of "solving" the risks of terrorist threats through iris scanners at airports is but one example.

By contrast, a vast and quickly growing literature has developed over the past two decades or so that focuses on the means by which technologies are socially constructed (Wajcman, 2002). Technologies in this conceptualization, are part of larger systems that encompass techno-scientific, economic, organizational (networks, alliances but also educational and other social systems), political, policy and societal institutions. We have conceptual handles and even now some excellent statistical measures of some of these. The challenge before us is to build on advances made to begin to conceptualize and capture the social environmental circumstances conducive to innovation potentiality rather than only focusing on existing partnership, alliances, networks, or clusters of innovators.

1. BEYOND MICRO-SOCIAL CONTEXTS

Not only do micro social contexts of organizations, firms and networks (Wellman, 2001) matter, as shown in other chapters in this volume, but also the macro-social contexts of risks, uncertainties, insecurities, education, health, environment, and community. This is not new, dare we say innovation, in thinking. Social theorist, Thorstein Veblen understood the phenomenon well as the 19th century was dying, in 1899. Industry, he argued, requires a spiritual aptitude to work, a commitment that goes to sustain the well-being of the community at large, and significantly, makes a caricature of the "economic" or "self-made man" since all experience, knowledge, technology and natural wealth are legacies of the community as a whole (Veblen, 1992[1899], p. 162; Veblen, 1939 [1915]). Lester Ward in the early 19th century took Veblen's insights one step further in asserting that individual success or genius, is the result of social opportunity rather than due to the efforts of any "self-made man" (1906, p. 252-261).

The fundamental importance of the larger social environment, the carapace in which firms, networks, alliances, individuals and socio-technical systems function, is brought home in recognition of the collective benefit of excellence in one person: "It is essential for all men [sic] that some individuals reach the stars" (Stewart, 1998, p. 124). But, of course, individuals do not reach the stars by a simple extension of their lone arms. It takes a village and a social structure to enable that star to be reached. Firms, alliances and networks depend on the entire social infrastructure to reach their "stars," most notably a highly educated workforce and enlightened consumers. That is the elusive social environment of innovation potentiality.

In development, adoption, uses and impacts, innovations are shaped by, as well as shape, social relations. Indeed, some scholars argue that scientists themselves are no more than social networks, including face to face or virtual connections with other scientists, with policy, social and economic environments, as well as with non-human resources such as semi-conductors, computers, laboratories, notebooks, statistics, and so on. This argument is compellingly made by Latour (1988) with respect to Louis Pasteur who, Latour argues, exists only as a network. Callon and Law (1997, p. 170) suggest that:

> ...the sociology of science and technology makes this argument. Entities –human, non-human, and textual – aren't solid. They aren't discrete, or clearly separated from their context. They don't have well-established boundaries. They aren't, as the jargon puts it, distinct subjects and objects. Instead, they are sets of relations, for instance, in the form of networks.

In more concrete terms, the processes of innovation can be observed and documented long after the product leaves the laboratory. Take the example of the microwave, which emanates from military radar technology. When microwaves first were marketed to the public, they were thought of as devices like VCRs that would sit next to a television for men, largely single men, to reheat prepared food. They were "brown goods" sold with televisions, video recorders, etc. for leisure and entertainment. But, innovative as we are in society, the technology traveled, in this case, to women, who appropriated the microwave as a "white good" (Wajcman, 2002). Interacting with the social realm, product innovation continued and the marketers took direction and achieved success not as a result of the product innovation itself, but because of the process of innovation which occurred in the realm outside the firm, in the realm of society at large.

Expanding on these insights, it seems timely perhaps in light of both the intellectual base we now have, and the strong policy interest in developing official statistics on innovation (Industry Canada, 2002; Human Resources Development Canada, 2002), to move beyond the micro-social contexts of innovation partnerships and alliances, and into the realm of social environments writ large. Innovation linkages draw on and connect with an educated work force, basic and science literacy, and with societal and economic infrastructure at all levels. Innovation potentiality in social environments can be seen as the larger system which encompasses smaller, micro-systems of innovation alliances and networks, nurturing them or detracting from their innovative potential. The social capital/social cohesion literature leads to an understanding that what society does, or does not do, matters greatly to achievements, including the capacity to innovate, both individual and collective in every realm of human endeavor. The concept of social capital leads to the realization that input from human beings matter, human beings in which we invest social resources such as education, self-esteem, the virtue of excellence, and perhaps most importantly, the idea that one's contributions matter not only to oneself but to the society.

Drawing on this burgeoning literature, we find that social capital is trust that flows through social networks (Coleman, 1990 cited in Rice and Prince, p. 210). Innovation is premised on such flows of trust and the existence of networks through which trust can flow. Many specific examples of how this works are provided by other chapters in this volume.

Science, like air, is best when flowing freely and freshly without contaminants or obstructions. And yet, science, knowledge and innovation are increasingly being rounded up by policy makers who, ostensibly in the public interest, seek to harness innovation and its fruits for economic and social benefit. The promised benefits of innovating are substantial and wide-ranging (see, for example, Industry Canada, 2002; Human Resources Development Canada, 2002). Not all policy initiatives, however, set their eyes on so glorious goals. An example of the extremes to which policy on science can go is found in the aftermath of September 11th. "In the blessed name of national security, and trumpeting the anti-terrorism anthem…Western governments are … moving firmly in the direction of information clampdown…Editors of scholarly science journals are being pressured not to publish anything that might be construed as aiding terrorists – regardless of the negative impact this will surely have on scientific advancement" (Posner, 2002, p. B15). Science does not neatly fit political categories and knowledge, by its nature, migrates. Borders and boundaries have not been able to keep most knowledge contained, although not for want of trying. And policy has not had a record of unmitigated success in containing or harnessing science to its own ends.

And yet, policy, imbued as it always is with contradictions, refers often to the social environmental factors that are conducive to innovation. For example, Industry Canada (2002) and Human Resources Canada (2002) policy papers on innovation, suggest that government can make a difference in promoting innovation and outline how. The "innovation environment" is seen as "the climate created by government stewardship regimes that protect the public interest and encourage and reward innovation" (Industry Canada, 2002, p. 62). "A truly world-class innovation environment...," it continues, "recognizes that the public interest must be protected." Public interest is in "health, environment, safety, privacy and consumer rights." Policy acknowledges that social environments are central to innovation. The challenge is to begin to capture these in statistical systems.

2. THE INNOVATION/SOCIAL RELATIONS WALTZ

It is not only that networks matter to innovation, but social relations and social structure shape innovation and subsequent potentiality for social relations that are conducive to further innovation. As well, innovation shapes social relations and networks, not only in processes, but also in perception and potentiality. Organizational and social innovations are key components of the process of knowledge advancement and of ongoing social change. Much evidence of how this works is presented in other chapters in this volume. Perception is crucial since firms, networks, alliances or societies can be either encouraging or discouraging of innovation by structures, attitudes and ways of interacting.

Sociological literatures on industrialism are replete with evidence of how complex social relations in the workplace work to the detriment or not of production and worker responsibility. The concept of commodity fetishism arose from an attempted denial, or wishful oversight, of these complex social relations existent behind commodities and their production. Branding of commodities in this post-modern era takes this fetishism one step further, in releasing products from their sources of production as they come to symbolize something very different and more abstract. The social division of labor behind commodities gets lost, and with it, its power to make change, to innovate.

The industrial democracy movement began to move away from Fordist preoccupations with efficiencies and back to personal pride and work group responsibility for commodity production. This maximizes investment in product creation, and reveals to both employers and workers that individual input and creativity is valued and visible in product production. The potential for contributing new ideas to industrial production is enhanced by the prevailing social relations in the workplace (Brown and Lauder, 2001). Precisely this was found to differentiate Statistics Canada from other federal government departments in a recent study of public sector departments. Not coincidentally, Statistics Canada has the best record both nationally amongst government departments, and internationally amongst statistical agencies for both innovation and for worker satisfaction. Innovation may be an oxymoron only some of the time in the public sector (to borrow and play on Earl, 2002).

The social capital/social cohesion literatures are beginning to reveal that economic and political benefits are not antithetical with social progress (see Bernard, 1999; Osberg, 2001). In fact, there is ample and growing evidence that social investments in health, for example, have direct and immediate economic benefits (Galt, 2002; KMPG, 2002). "Publicly sponsored health care [in Canada] has been a real source of competitive advantage, particularly over the United States," (Galt, 2002:B10). An international comparative study of competitive advantage finds that Canada is the overall cost leader with a 14.5% cost advantage over the United States (KPMG, 2002). Canada has cost advantages in every sector, largely as a function of publicly sponsored health care. Canada's largest cost advantage occurs in sectors where innovation is expected or normative, such as software (cost advantage of 22.3%), research and development (30.6%), and corporate services (19.7%). In all of these, the U.K. is second, with the U.S. being near the bottom, except in corporate services where the U.S. ranks fourth. The question arising in this chapter is the degree to which public infrastructure enables innovation potentiality, and whether this can be captured in statistical systems. And of course, what else may matter in social environments to innovation potentiality.

3. FRAMING STATISTICAL CAPTURE IN GOVERNMENTALITY

The challenges of statistical capture of the processes of science and technology are staggering, as those of us who have been closely involved can attest. Challenges peel back like in layers like an onion. A key challenge is that official statistics, although neutral, do not come from the ether. There is no view from nowhere, to paraphrase a socio-historical study of censuses in Canada (Curtis, 2001, p. 30). "Some constructed accounts of the world are more authoritative than others because they can enlist powerful allies in their support" (*ibid*). Others are authoritative because they confirm our hunches, and still others for the opposite reasons, because they disconfirm, with the authority of officialdom, what we thought we knew.

Science in the making is messy, haphazard, uncertain and characterized by conflict and debate (Latour, 1987). This is no less the case for innovation and its alliances, networks and partnerships. "Made science" is black-boxed, its dependency on particular modalities of investigation and enlistment of supporters' claims begin to be taken for granted. Students who are taught science by rote, and many still are unfortunately, are sometimes shocked with the discovery, later in their studies, that science is an adventure into the unknown, with only our imaginations and our integrity to guide us.

Just as census making (Curtis, 2001, p. 33), and official counting of crimes (Haggerty, 2001) configure social relations in keeping with political and cultural obligations and interests, so must it be with counting and capturing alliances, networks and partnerships in innovation. Of necessity, some actors and activities are highlighted for data capture while others are excluded. We capture innovation where the light on innovation is best, and leave the dark corners or the hot sunny areas for later. We focus, for example, on existing networks or successful alliances more than those yet to form or those that failed or collapsed.

> When you can measure what you are speaking about and express it in numbers you know something about it, but when you cannot express it in numbers, your knowledge of it is of a meager and unsatisfactory kind; it may be the beginning of knowledge, but you have scarcely in your thoughts, advanced to the state of *science*, what the matter may be. (Lord Kelvin quoted in Thomson, 1889, p. 73, italics in original.)

Making innovation count, to borrow from Haggerty's turn of phrase in his book, *Making Crime Count* (2001), involves transforming innovation and the processes in which it occurs, into something that can be counted and made interpretatively meaningful. Statistics are bathed in the social, as Sir Leon Radzinowicz' comment about French bathing suits suggests (and I paraphrase it slightly for this context): Innovation statistics are like French bathing suits; what they reveal is highly suggestive, but what they hide is vital!

Statistical knowledge plays a vital, and increasing, role in liberal governance (Foucault, 1991). Before any particular subject can be governed, its distinctive form, inclinations and tendencies must be known. Surprisingly little attention in the social sciences has been directed to the institutional production (capture) of official statistics (Haggerty, 2001, p. 5; Curtis, 2001). The reification effects of official statistics has had even less attention, and yet it is strongly apparent, particularly so when answers are eagerly awaited to difficult questions of governance or policy. When the big question is whether there is a recession in Canada, we look to official statistics: "Come Thursday, Statistics Canada will fill in the remaining holes in the puzzle and frame the whole thing with the all-important edges and corners that will reveal the complex picture" (Little, 2002, p. B7). Official statistics not only count, but paint a picture in its complexity, a picture on which people and policy-makers as social actors, can act.

4. TOWARD THE CAPTURE OF INNOVATION POTENTIALITY

All this said, where and how do we begin the process of capturing in official statistics the elusive concept of innovation potentiality? Two answers are suggested, and they may appear contradictory. First, we have already moved a considerable way forward along some dimensions in the capture which will be shown by way of an example. Second, we start with baby steps, in much the same way as we did in the capture of technological and scientific impacts and outcomes on society (Statistics Canada, 1998). These contradictory answers can be reconciled in the observation that we know much (or at least some aspects) of what we need to know, but we have not yet assembled the puzzle pieces into anything coherent. That was, in part, the case when we developed the framework for the capture of statistics on science and technology (Statistics Canada, 1998).

The example, an incomplete but compelling model of how we might capture innovation potentiality, comes from careful analyses of data not collected *per se* under the purview of innovation or science and technology statistics. Bordt *et al.* (2001a; 2001b; 2001c) take a detailed look at the development and acquisition of science and technology skills from elementary school through to labor force participation. The study is presented in three movements, much like a symphony, relying on diverse data sources such as graduate surveys, censuses and international comparative data from OECD. The findings reveal much about the carapace of innovation potentiality, and provide a model of how we might proceed to mine existing data to capture dimensions of the wider social environment that impinge on innovation potentiality. The findings are surprising and add punch to the importance of knowing not only how innovation occurs in those micro-contexts into which we have cast our torchlights, i.e. firms, networks and alliances, but what occurs in macro-social environments with respect to innovation potentiality.

Interest in science and technology is found to decline among Canadian students early, between Grades 4 and 8, and then continues to drop during high school (Bordt *et al.*, 2001c, p. 9). Even students who do well in sciences and maths and those who believe these subjects are important for careers, seem unwilling to pursue these subjects. That this dropping phenomenon is found to be more pronounced in Canada and the U.S. than in other OECD countries, suggests that "socio-cultural factors… could be influential" (Bordt *et al.*, 2001c, p. 10). What factors, they understandably cannot say, because it goes beyond the scope of the data at hand. This study, however, points toward exactly the kind of factor we would wish to pursue to capture innovation potentiality. Such factors as "economic climate, unemployment rate and proportion of immigrants in the population may be important in explaining national differences in performance" (Bordt *et al.*, 2001, p. 10). Evidence from other studies suggests that "Two years after graduation, more than half of all master's graduates...felt that they were overqualified for their job" (Bordt *et al.*, 2001c, p. 10). This is innovation potentiality stifled. Why, we ask, would this occur? And what kinds of new data and new analytical approaches to existing data might be utilized to find answers. The authors add, enticingly, that "complementary skills acquired during postsecondary education (or by some other means, such as self-directed learning) can be more important than those relating directly to the main field" (Bordt *et al.*, 2001c, p. 11). So, it is outside the educational system per se that the biggest benefits are gained.

The analysis continues to shed light on innovation potentiality, light that may matter to policy and to enhanced public understanding of what kind of macro social environmental conditions may be conducive to innovation. "[G]eneral macro-economic and social context (such as the regional unemployment rate and proportion of immigrants) may be as important as social spending in influencing outcomes" (Bordt *et al.*, 2001a, p. 20). Of course, the authors state, with prudence and accuracy, that such factors go beyond the scope of the current study. If, however, we are confident in what we cannot otherwise know, and confident in the truths that existing official statistics provide, then it may be time to jump beyond into these previously "unassembled for the purpose" data that could reveal more of innovation potentiality.

What then are the baby steps that might be taken in the direction of capturing innovation potentiality? Relying on the previous approach to developing official statistics on science and technology, the direction of the baby steps may be the same, except here the challenges may be both larger and smaller. They are larger in that the data to be captured are wider and much more diverse. The challenges, however are smaller in three ways: first, deep conceptualizations exist and could provide guideposts; and second, the process can benefit from the previous experience of developing a cohesive framework for an information system for science and technology (see Statistics Canada, 1998); and third, the model of the population health framework which has transformed the way health and health policy are seen (ably summarized by Evans and Stoddart, 1994) can be a guiding light for this endeavor.

The steps to be followed are two, and they can proceed simultaneously. Some clear-headed "blue sky" thinking should proceed about the components of innovation potentiality with the goal of developing a cohesive intellectual framework. At the same time, existing official statistics known to be relevant on education (as seen in the example explored above), on health (here the population health model may be directly relevant) and on social inclusion, might be rounded up and closely scrutinized for their capacity to shed light on innovation potentiality in social environments. A grid might be developed where absences of needed data are apparent.

5. CONCLUSION

Rather than critiquing existing approaches to the capture of official statistics on innovation, this chapter argues that we could build on those approaches and experiences, in both theory and data, to confidently leap to the next level. Our capture of science and technology impact and outcomes, although admittedly a project in the making, has led to a productive convergence of knowledge. We know enough theoretically and now empirically, that we no longer with equanimity can fall back into the "beyond the scope" or "there are no data" arguments. The knowledge base exists for moving into capturing the macro-social forces and factors that condition innovation potentiality in society. To the extent that data exist already in official statistics, perhaps the time has come to accept the challenge of developing a cohesive framework of innovation (or build with small blocks the beginning pieces of the explanatory picture). Policy could benefit as could our insights into what it takes to be an innovative country (see Fagan, 2002). The lessons from the population health model are clear -- both understanding and policy were impeded by reliance on misleading frameworks of health determinants. Appropriate conceptualization is necessary, but of course not sufficient, to serious attention to innovation and policy-making with respect to innovation.

Chapter Fourteen
POLICY NETWORKS IN ADAPTIVE INNOVATION SYSTEMS

John de la Mothe
Canada Research Chair in Innovation Strategy,
PRIME, University of Ottawa, &
Visiting Professor of Science and Technology Studies,
University College London

> What is a network? It is a certain form of multifaceted inter-organizational relationship through which new valuable knowledge is generated. Ken-Ichi Imai (1989)
>
> Firms may have to collaborate and network more....and strengthen their links to the science and engineering base, sharing equipment and often people. Department of Trade and Industry, U.K. (1998)

1. INTRODUCTION - THE IMPORTANCE OF NETWORKED POLICY

Public policy-making is, by definition, the result of network interaction. And in our complex interactive world which Ralf Dahrendorf typified as 'the Age of Schumpeter, the Age of Innovation',[1] this notion of policy networks in support of innovation has taken on renewed, intensified, meaning. After all, as most governments understand, innovation is intimately related with economic growth, productivity, competition and improved standards of living. (OECD, 2001c)

Innovation policies are thus focally concerned with stimulating, guiding and monitoring knowledge-based activities. They are based fundamentally on the performance of scientific research, technological development and diffusion, but can only be understood within the context of performance and application within organizations, regions and nations. These contexts differ across countries and across industries. The goals of innovation policies are principally economic, although they have important social aspects and impacts. The design and deployment of innovation policies take place within a series of complex interactions and networks between agents and institutions. Moreover, with the increased pace of interaction across borders and spaces and the interdependencies between agents it can be said that innovation policies and networks are growing in importance and are becoming more complex. Meeting the challenges of a dynamic and interdependent economy will therefore require substantial advances by policy makers in understanding of how our technological, innovation and policy systems interact. These systems include networks of knowledge producers (such as researchers and research groups), knowledge users and appliers (such as firms), knowledge regulators (such as intellectual property agencies), knowledge diffusers and infrastructures, and knowledge funders (such as granting agencies). Thus the governance challenge of innovation policies is real, but the response is to move towards adaptive policy frameworks.

Towards this end, this chapter will discuss networks in innovation systems, notions of networking in the policy process, and a few real examples of networked policy responses within an innovation system.

2. NETWORKS AND INNOVATION

The concept of networks allows us to give specific meaning of both the idea and the mechanism of an 'innovation system' in which technology transfer and knowledge/information flows are key. Indeed, the analytic notion of networks allows us to more accurately represent the structure of a social system that is constituted by a set of actors and by the linkages or interactions that connect them. Two important aspects of networks are their stability and their efficacy. Efficacy here implies the achievement of network goals. This is very different from the idea of efficiency. There are numerous analytic and policy issues embedded in this statement, but if networks are an important ingredient to a nation's economic performance then their creation and maintenance of vibrant networks are an important and legitimate policy concern. (Edquist, 1997)

Ever since B.Ä. Lundvall re-introduced the 19th century work of List through innovation systems to scholars and policy makers between 1988 and 1992, it has become *de rigeur* to defer to the observation that interactive learning and collective entrepreneurship are fundamental to the processes of innovation. (Lundvall, 1992) This observation highlights the networked reality of innovation - of getting ideas to market - rather than the linear models of innovation. In network framework, innovation must be seen as an iterative, cumulative and cooperative activity, not the result of individual decisions. (de la Mothe and Paquet, 1998b) In other words, firms and institutions never innovate in a vacuum and are seldom capable of innovating independently. Let us illustrate with a famous example and a few useful quotations.

The introduction in 1977 of the Apple II microcomputer was in many ways a landmark. It was not the first microcomputer to be sold, but it was on of the first to give the PC industry serious momentum and attention. Indeed, as Swann and Prevezer (1998) point out, it was largely in response to the success of the Apple II that IBM entered the PC market in 1981. The interest of this case is that Apple was success even though they enjoyed very little vertical integration at that stage. "The product was designed to be an open system so that it could draw on the large crop of add-on boards, software, and peripherals that started to appear from third party suppliers." (p.4) In the early days, Apple depended on external suppliers for numerous components. Apple saw their business as designing, educating and marketing, and the production or assembly of the hardware could be done through sub-contracting. Of course, as Rothwell and Zegveld highlighted in 1984, given that no firm can 'know it all' and that competition implicitly requires a continuous 'search' or scanning function, by dedicated sub-contracting firms essentially extended the supply- and value-chains into a networked innovation system. For Apple, this strategy succeeded because of their networks in Silicon Valley. Not only did this *network strategy* succeed for Apple, but it also became the strategy that IBM decided to emulate in order to enter and compete in the PC market.

This is but one example. Bob Rycroft and Don Kash, (1999) in their book *The Complexity Challenge: Technological Innovation for the 21st Century*, amplify this case by making the following series of observations:

> Although it is popular to think of the corporate chief executive officer as a 'captain of industry' who, like a ship's captain, commands the behavior that keeps the ship on course, given the ubiquity of complexity, leaders have come to accept their limited ability to control as a fact of life. They have become comfort-able operating with a more general understanding. Whether they are company presidents, technological entrepreneurs, program managers or design engineers, the message is the same, *success occurs absent of detailed understanding by individuals*. (emphasis added) (p. 3)

[Here are some relevant quotes from senior executives.]

- Hitachi is too complicated for anyone to run it. As president, all I can do is think and talk with people about the future and then communicate those thoughts to others in the company.

- Intel operates on the Noyce principle of minimum information. One guesses what the answer to the problem is and goes as far as one can in a heuristic way. Thus, rather than mount research efforts aimed at truly understanding problems and producing publishable technological solutions, Intel tries to get by with as little information as possible.

- There was no way Hewlett-Packard and Philips could know in advance what our relative contributions to the project would be, so we agreed to decide how the profits would be split after the technology was on the market.

- The Japanese supplier delivered a 97% finished product that had been through repeated prototype tests before we furnished them the turbine's specifications.

> What is happening? How do corporations, government agencies, and other organizations achieve success despite the limited understanding of those who lead them? *The general answer is that success depends heavily on continuously adaptive organizational networks that know how to do more than any individual can understand in detail.* (emphasis added) These networks manifest patterns that are very different from those of traditional organizations.... *In the realm of complex technologies, organizational networks have blurred the line between the public and the private sectors and between company strategies and public policies.* (emphasis in the original)

Chris Freeman and Francisco Louçã also embrace the importance of complexity and networks and generalize the economic point by noting that:

> ...no structure can be postulated to represent the fixed form of the social process, but a structure does indeed exist, although it is evolving. Complexity, in this case, refers not only to the feedback processes, but also and most importantly to the change of the structure during the process. This is evolution, and this is why history cannot use either static equilibrium theories or dynamic models based on reversibility, but must resort to processes that have the characteristics of irreversibility, non-ergodicity, and path-dependence. (Freeman and Louçã, 2001, p. 41)

Again, such observations are portentous for policy makers.

If one were to further examine the examples of industrial innovation afforded us by Chris Freeman (1978), David Landes (1969), and Alfred Chandler (1977), then one would see a long history of interdependency between firms in their quest for innovation, progress and 'creative destruction'. What they would also find is a growing dependency on the emerging factor of production known broadly as 'knowledge' - referring to science and technology. Indeed, upon closer inspection one would discover that industries and technologies differ significantly in their approach to accessing new productive knowledge. (Remember that Alfred North

Whitehead, in his *Science in the Modern World* [1925] said that 'the greatest invention of the 19th century was the invention of invention itself'.)

In biotechnology, for example, - especially biopharmaceuticals - the investment requirements, the risk lead-time, the intellectual property protection, and the access to primary university research needed - all popularly 'captured' by the so-called 'S-Curve' - are higher than in the general fields of information and communication technologies.[2] So in briefly highlighting the fact that firms do network and need to network in order to compete in what can be described as 'innovation systems', let us now turn to some internal dynamics of networks.

3. NETWORKS IN PRINCIPLE

The fundamental uncertainties that are central to the innovation process lead to a series of activities involving search, experimentation and what firms would see as satisficing behavior. In much of the innovation literature, this bundle of activities is referred to as a learning process. The required information for successful innovation is typically unknown or unpredictable. Cooperation with others is therefore often required. But how should this be structured?

In complex tasks, the expertise and competence of others is often needed. As David Teece and David Mowery, for example, have long articulated, firms must learn to choose which agents to cooperate with - resulting in what Teece calls 'cooperative competition' (Teece, 1990) - through an array of strategic alliances, mergers and acquisitions, contract research, supply chain management, and so on. But this does not so easily inform the conduct of policy making in innovation systems.

Networks are an intermediate form of organization between hierarchies and markets. Their essential function is the exchange of knowledge, information, technology and value. In economics, exchange functions are often reduced to market mechanisms that like to feature an imaginary set of perfect information typified by 'prices' and rationale (that is, maximizing) actors. But when transactions involve complex competencies, the market does not work at all well. Networks are required.

Networks are informal. Their boundaries are often difficult to define as actors will ebb and flow, pursuing varying needs and offering various competencies. Network boundaries therefore are typically in a form of flux.

Gelsing (1989) identifies four types of industrial networks: user-supplier linkages, production chains, production *filiéres*, and knowledge networks. The difference between production *filiéres* and knowledge networks focuses on the differences between content - material resources versus competence resources. While some of the information transferred in knowledge networks can be marketed in the form of market analyses or training manuals, most of the information is transferred through informal exchanges. Typically, such exchanges rely on the interface between technicians, scientists, engineers, experts and practitioners. Also too, such exchanges typically focus on non-standardized problems or opportunities. (We can infer much for policy from these typifications.)

Eric von Hippel (1990) has noted that in such non-standardized issue areas, the shape of networks can be explained in terms of its 'stickiness'. Stickiness may be the result of proprietary information, it may be the result of tacit knowledge or it may be a function of incompatibilities between the transmitter and the receptor. As von Hippel puts it:

> ...at one end of the scale are data with zero stick-iness that can be replicated and diffused at zero cost. At the other end are data that are so sticky that they cannot be replicated or diffused at any price. (p.1-2)

This observation has important implications for policy making.

Finally, two things that networks require in order to function are an entrepreneurial locus and a capacity to develop critical mass. In small and medium sized firms, regions or nations, the network approach can in fact act as a compensatory vehicle for the development of critical mass. In many ways this reveals the underlying logic of network approaches in countries like Canada.[3]

4. NETWORKS IN POLICY

The policy and public administration literatures on networks are less helpful. While the notion of networks is prevalent in innovation studies, so too do we find that policy studies and public administration scholars have dealt with networks. However, the two have yet to be sufficiently linked. Indeed, Leslie A. Pal overstates when he writes that "there has been a harmonic convergence around the ideas of policy networks". (Pal, 2001, p. 236) In point of fact, any cursory review of some major public administration texts - such as Kerneghan and Siegel (2001) - will reveal not a single reference to innovation, science, technology, research or knowledge in the widely accepted sense despite the fact that these are widely accepted by governments as being key factors of production that are dramatically challenging governance and policy making while driving globalization, economic growth and employment patterns. Indeed, to illustrate the wide variants of meaning around innovation networks between innovation scholars and policy researchers, Leslie Pal 'defines' innovation networks as "networks that bind together public administrators, managers, consultants, and academics to ensure that examples of public sector innovation in program delivery are as widely available as possible". (Pal, 2001, p. 224) Clearly this definition has nothing to do with innovation as defined by Schumpeter. Nor are the mechanics of network formation usefully discussed through the term "bind together". Whereas Carlsson's 'technological systems', Eliasson's 'competence blocs', Lundvall's 'innovation system' and even Dosi's 'badly behaved dynamics'[4] aspire (with some success) to investigate the mechanisms and dynamics of network formation and maintenance, much of the policy literature seems remote from the realities of adaptive and evidence-based policy making.

Nonetheless, some policy scholars such as William Coleman and Anthony Perl (1997) have usefully reviewed the literature on policy networks. Since its appearance in the 1970s as an analytic term within the public administration and policy communities, 'policy networks' has come to mean "a set of informal and formal interactions between state and private actors who have different but interdependent interests". These actors operate within institutional settings and they come together in horizontal, relatively non-hierarchical, discussions or negotiations. The purpose of policy networks is to define problem areas and policy options.

An important aspect of this definition of policy networks is its signaling of a change in the way in which political decision-making has been understood. It eschews the standard notion of a clear separation of State and Society, and embraces the co-evolution and interaction of government-business relations. (de la Mothe and Paquet, 1996) Clearly this broad observation is complementary to the systems approach developed by Robert Jervis (1997) (system effects), Bo Carlsson (2000) (technological systems) and Gunnar Eliasson (2000) (competence blocs). But in the policy literature, much remains loitering at the supra-level of macro description. There is no theory of mechanics nor is there any normative imperative implied. But this is hardly surprising given the remaining sense that policy scholars are still searching for an agreed upon problem. For example, while U.S. scholars tend to discuss issue networks, Scandinavians interrogate corporate pluralism, British researchers tend towards policy communities, and international relations theorists prefer epistemic communities. (Coleman and Perl, 1997)

For Peter Haas (as an example, an epistemic community is a network of professionals with competence or expertise in an area. Although such a community may stem from differing backgrounds and disciplines, Haas holds that "they (1) have a shared set of normative and principled beliefs, which provide a value-based rationale for the social action of the community members; (2) shared causal beliefs, which are derived from their analysis of practices leading or contributing to a central set of problems in their domain and which then serve as the basis for elucidating the multiple linkages between possible policy actions and desired outcomes; (3) shared notions of validity - that is, intersubjective, internally defined criteria for weighing and validating knowledge in the domain of their expertise; and (4) a common policy enterprise - that is, a set of common practices associated with a set of problems to which their professional competence is directed". (Haas, 1992, p. 3) Haas and his colleagues then go on to 'deploy' their 'framework' (which is only macro-socially concerned with power and hegemony) to force such diverse issues as the Uruguay Round of trade talks, the protection of stratospheric ozone, international food aid, nuclear arms control, the management of whaling, and central banking into a 'model' that only says that 'smart people get together when they need to around an issue'. This is hardly useful for either scholarship or decision making as it doesn't help to explain how, when, and why networks form, and it does not help decision makers evaluate the issues around innovation (in our case) and what to do about it in each of our industrial and national contexts.

This broad dispersion of language and frameworks in the policy literatures is thus - quite the opposite of Pal's 'harmonic convergence' - more reminiscent of The Goon's (1957) BBC skit "Six Charlies in Search of an Author". Clearly, much policy and public administration research has failed to focus on how decisions get made and what (normatively) policy makers should do in a knowledge-based economy. Rarely if ever do they seem to venture into labs in order to raise their understanding of how science, technology or innovation 'gets done'. Rarely, if ever, do they examine whether their own assumptions regarding supposedly common principled beliefs, shared causal beliefs, shared notions of validity, or views of common policy enterprises are valid, discernable or operational. And rarely do they define theoretically who is 'qualified' to join an epistemic community. (Indeed, it would be surprising to find 'an epistemic community' that didn't feature significant issue contestation in the Kuhnian (Kuhn, 1962) or Mertonian (Merton, 1973) sense. Instead, these approaches are stuck at either the galactic level of power and dominance or at the street-corner level where everything can be described but nothing can be learned.

5. POLICY NETWORKS IN REALITY

Governments have done better. As I have argued elsewhere, government policy makers have in fact been highly adaptive.[5] Nations, from Norway, Ireland, Germany and Sweden to Australia, Canada and South Africa, have used the 'innovation system' as an organizing image or framework. Those countries that do use this set of images and understandings have shown that innovation policy can be an exercise in adaptation within the demanding environment of globalization, challenges to governance and dispersion of political power, and the need for rapid responsiveness to events and new information given the penetration of information and communication technologies.

In Canada, where a variety of rich examples can be found, the need to network has been a policy focus for many years. Something of a motto within the government and industry associations has been "smart industries are connected industries". This is particularly relevant since Canada is a vast country with many dispersed institutions and a thinly distributed population of

30 million people with 85% living within 100 kilometers of the 49th parallel across a 3500 kilometer border. In today's environment, the government has aggressively recognized that linking strengths can create critical mass. In some ways the roots of this thinking can be traced to 1916 with the creation of the National Research Council (NRC) which, particularly since the 1930s, has evolved into a network of federal government laboratories in pursuit of the goal of nation building and regional economic development. (Dufour and de la Mothe, 1993) More deliberately in the network and innovation sense was the creation of the Industrial Research Assistance Program (IRAP) at the NRC in 1947 when Canada began to enter the ranks of the advanced industrial nations. But let's briefly describe a small sample of programs and policy adaptations that are network-oriented.

5.1. Industrial Research Assistance Program (IRAP)

The mandate of the IRAP is to stimulate networked innovation in Canada small and medium sized firms (i.e. under 500 personnel). To do so, IRAP staff provide access to technical and informational resources to support client activities. The IRAP Network is composed of more than 130 public and private research and technology based organizations. This network serves as an innovation infrastructure. IRAP adheres to the idea that

> ...companies that master the process of harnessing creativity to create new value in new ways earn financial returns. Wall Street [and here we should add Bay Street] increasingly places higher value on innovation than on virtually any other approach to generating growth. More than leadership change, more than mergers and acquisitions, more than renewed commitment to cost reduction, investors reward innovation.[6]

Clearly the IRAP Network embraces the idea that "to be successful, a firm must be innovative on all fronts - from R&D to business and human resource functions. The Canadian Innovation System includes various agencies.... That can assist SMEs with various aspects of innovation."[7]

Indeed, as they say, "when your innovation strategy goes beyond simply buying the latest equipment, but includes the adoption, adaptation or creation of new technology for the development of new products or production processes, IRAP can help you."[8]

An important element of IRAP is the idea of pre-commercialization assistance for enabling technologies. Enabling technologies in this context refer to aerospace and defense (including defense conversion) and environmental technologies. It focuses on advanced manufacturing and processing technologies, advanced materials, processes and application, applications of biotechnology and ICTs. Pre-commercialization assistance under IRAP includes the development of new or improved goods and services, as well as initial demonstration and pilot projects. Supportable activities include scale up activities; engineering to meet user requirements; manufacturing engineering; pilot facilities; and testing.

5.2. Pre-Competitive Applied Research Network (PRECARN)

Another innovative example of policies in support of networks is PRECARN. PRECARN Incorporated is a national consortium networking firms, research institutes and government partners working within the intelligent systems industry. Created in 1989, PRECARN helps bridge the 'innovation gap' between university and government research and their commercial application - again, taking to heart the definition of 'innovation' that was offered at the outset of this chapter. Within a 3rd phase renewed commitment from the government in excess of $35 million until 2005, PRECARN seeks to foster growth in SMEs by:

- Improving productivity and competitiveness;
- Supporting the development of new, commercially viable, technologies;
- Encouraging the diffusion and market exploitation of these new technologies by Canadian firms, and
- Developing - and sustaining - a cadre of highly qualified personnel in the intelligent systems industry.

This last component, which is present in all the network-oriented programs mentioned here and extant in the Canadian government system, is key to approaching its goal of moving up in the international ranks of knowledge producers and users.

5.3. Networks of Centres of Excellence (NCE)

The NCE Program fosters the creation of partnerships between all three economic sectors. Networks that are funded by the program are designed to develop Canada's economy and quality of life. Three federal granting agencies - the Canadian Institutes of Health Research, the Natural Sciences and Engineering Research Council and the Social Sciences and Humanities Research Council - plus Industry Canada have combined to support and oversee the NCE initiative. The Networks build on a strong base of university research and are concerned with stimulating innovation, training and retaining talent, and encouraging the transfer of technology and its diffusion.

Table 1. Currently Funded Networks of Centres of Excellence:

Health, Human Development and Biotechnology
Canadian Arthritis Network
Canadian Bacterial Diseases Network
Canadian Genetic Diseases Network
Canadian Language and Literacy Research Network
Canadian Network for Vaccines and Immunotherapeutics
Canadian Stroke Network
Health Evidence Application and Linkage Network
Protein Engineering Networks of Centres of Excellence
Stem Cell Network
Information and Communications Technology
Canadian Institute for Photonics Innovations
Canadian Institute for Telecommunications Research
Geomatics for Informed Decisions Network
Institute for Robotics and Intelligent Systems
Mathematics of Information Technology and Complex Systems
Micronet - Microelectronic Devices, Circuit and Systems
Natural Resources
AquaNet - Network in Aquaculture
Canadian Water Network
Mechanical Wood-Pulps Network
Sustainable Forest Management Network
Engineering and Manufacturing
AUTO 21 - The Automobile of the 21st Century
Intelligent Sensing for Innovation Structures

5.4. Canadian Technology Network (CTN)

The policy goal of the CTN is to connect (i.e. network) people with expertise. The Canadian Technology Network links federal and provincial government labs and agencies, universities, community colleges, industry associations, technology centers and economic development agencies. Together, as a network, they provide firms with access to needed expertise, advice and information about technology-related challenges.

The mission of the CTN is to provide integrated and accessible pathways to services for small and medium sized business using technology. The first point of contact for firms is through a network of advisors.

5.5. Community-University Research Alliances (CURA)

Of course, let us not forget (as we often do) that social innovation is extremely important in the completion of the innovation cycle. Marketing, getting ideas diffused, the processes of technology transfer, and so on are more social and community processes than they are strictly technical issues. In this vein, and in keeping with the networking and adaptive nature of policy in innovation systems, the Social Sciences and Humanities Research Council announced - on February 5th 2001 - the Community-University Research Alliances. The $9 million program promotes active linkages between community groups and universities. The various projects call on the strengths of the community partners.

6. CONCLUSION

To conclude, policy networks are critical to adaptive innovation systems. To be sure, policy making is about networks and policy networks, even though policies in a fast moving interactive environment are sticky. That is to say that the shape of policy responses to contemporary knowledge-based issues still retain degrees of political positioning, interest lobbying, and bureaucratic territoriality. But policy makers have shown themselves to be - not without contestation - rather adept at contriving new and alternative forms of service and program delivery, even in the intangible and difficult realm of innovation policy. While innovation scholars have tried - and continue to try - to address the real issue of the innovation process, the world of policy scholarship needs to better appropriate this research in order to better inform decision makers and to provide decision makers with the tools that they need.

PART V

CONCLUSION

Chapter Fifteen
CONCLUSION

John de la Mothe
Canada Research Chair in Innovation Strategy,
PRIME, University of Ottawa, &
Visiting Professor of Science and Technology Studies,
University College London

and

Albert N. Link
Department of Economics
University of North Carolina at Greensboro

One theme that is present in each of the chapters in this research volume is that alliances, networks, and partnerships are important. They are important to the innovation process because they represent innovative strategic arrangements that both enhance the process as well as foster mechanisms to internalize knowledge spillovers. They are important to policy makers as well since they describe current trends in innovative activity and thus foretell related patterns in innovation-related investments that both private-sector organizations (e.g., firms) and public-sector organizations (e.g., universities and government agencies) will be making. The latter are especially important both as an indicator of flows into the stock of technological knowledge – an important current policy issue – and as a barometer to the changing nature of the boundaries of organizations – a burgeoning policy and governance issue.

From the private-sector side (e.g., firms), and that is the side from which there appears in this volume to be the greatest agreement that there is the least understanding about the nature – scale as well as scope – of alliances, networks, and partnerships, a recounting of aspects of Ronald Coase's theory of the nature of the firm is a useful starting point. It was Coase, in his famous *Economica* article published in 1937, who clearly articulated the firm as an economic entirety that exists to minimize transaction costs. A firm will exist by internalizing those activities for which the costs associated with market transactions are greater than the costs associated with internal activity.

Firms have long been understood to invest in in-house research and development (R&D) activity because the costs associated with outsourcing are perceived to be too great. These perceived differences in transactions costs are related to differences in the ability of firms to appropriate the knowledge that results from R&D when conducted internally as opposed to externally. Many extrapolate from their firm calculus to conclude that there is a market failure because of appropriability issues and thus there is an under investment, for a social perspective, in total private-sector R&D investments.

Each of the chapters in this volume has touched on, in a unique, important and provocative way, the increased effective use of alliances, networks, and partnerships to achieve their innovation strategies. As such, each is implicitly arguing that the boundaries of the firm have changed over time, or in the spirit of Coase, the transaction costs of external R&D-related activity has decreased. This interpretation of this trend would not be complete without a hypothesis as to why that is the case. While the authors of this volume did not address this point, and appropriately so given the specific topics of their individual chapters, we appropriately, at least in our opinion, do so in this Conclusion because this is a collective theme throughout the volume. Also, by doing so we hope that we also set forth a policy and a research agenda for others to expand upon.

Technological competition has increasingly become global in scope and related technology life cycles have shortened; firms have correctly responded to this new order by implementing multifaceted innovation strategies that reflect a new philosophy about the interdependence of competing firms. Speed in innovation is increasingly becoming the strategic benchmark upon which competitive survival will be benchmarked. As such, firms are partnering with other firms, organization, and institutions in an effort to survive, and are thus trading off a loss in appropriability for timing. Therefore, our hypothesis is that this new technology-based competitive order is changing the boundaries of the firm. And, as these boundaries change, the role of the public sector will become even more important.

While the public sector has always been important as a partner in providing direct and indirect resources in support of innovation, if technology policy is to succeed in the sense of assisting domestic industries to successfully compete in global markets, then an important policy level will become government's ability to successfully facilitate alliances and networks and to partner with private-sector firms in a timely manner.

As we increasingly move into what we've called an era of 'alliance capitalism', the observations and insights into the innovation process through networks, alliances and partnerships will grow in value. As such, we hope that this research volume will enable researchers, statisticians and decision makers to improve their analytic radars and to open up new avenues of analysis and action.

NOTES

Chapter Two

[1] Much of the economics literature on these topics is reviewed in Hagedoorn, Link, and Vonortas (2000) and in Jankowski, Link, and Vonortas (2001).

[2] See Hébert and Link (1988) for a complete history of the concept of the entrepreneur.

[3] And, perception of opportunity may be enhanced through public policies that emphasize particular technological directions.

[4] See Link (1987) for a review of these frameworks.

[5] See Link (1999).

[6] See Coburn (1995).

[7] See Audretsch, Link, and Scott (2002).

[8] See GAO (1994).

[9] See Hagedoorn, Link, and Vonortas (2000).

Chapter Three

[1] The significance of (Langrish, *et al.*, 1972) as a departure point is highlighted by the emphasis on the individual inventor in (Jewkes, J., Sawers, D., Stillerman, R., 1969) and the criticism of the first edition of this latter work by (Freeman, 1967) who noted the potential bias of the sample as well as suggesting several of the main ideas of what he would later call the national system of innovation.

[2] For review articles, see (Cohen and Levin, 1989; Dodgson and Rothwell, 1994; Stoneman, 1995)

[3] Within this structure, the publicly supported research of member states have continued to focus on domestic institutions.

[4] This is not an exclusive reason for inter-organizational agreements. Such agreements may also be made to support regulatory compliance, establish industrial standards, or engage in tacit collusion for the purposes of achieving anti-competitive outcomes. The focus of the current chapter is, however, on collaborative arrangements motivated by innovation.

[5] The argument advanced by (Nelson, 1959) and (Arrow, 1962) that problems in appropriating the full value of scientific information would cause private agents to under-invest in its creation. From this perspective, the creation of research 'clubs' producing 'quasi-public' goods is a compromise between achieving the full social value of information production and the social costs of producing this information solely through the use of public resources.

[6] This anecdote was devised by Nathan Rosenberg.

Chapter Four

[1] For more details on the methodology of the two surveys see: Schaan, S. and Anderson, F. (2001) and McNiven, C. (2001)

[2] The size categories used for the Survey of Innovation 1999 are: Small: 20-99 employees; Medium: 100-249 employees; and Large: 250 and more employees.

[3] The manufacturing sectors were broken down into 31 industry grouping based on 3-digit or 4-digit NAICS codes (North American Industrial Classification System). In some cases, 4-digit codes were combined to produce the industry grouping. For more detail on these categories and for the corresponding NAICS codes see: Schaan and Anderson (2001), pp.39-40.

[4] Martin, S. and Scott, J.T. (2000). "The Nature of Innovation Market Failure and the Design of Public Support for Private Innovation", *Research* Policy, Vol. 29, pp. 437-447. Their taxonomy is based on previous work done by and Pavitt, K. (1984). "Sectoral Patterns of Technical Change: Towards a Taxonomy and a Theory", *Research Policy*, Vol. 13, pp. 343-373 and Robson, M., Townsend, J. and Pavitt, K. (1988). "Sectoral Patterns of Production and Use of Innovation in UK: 1945-1983", *Research Policy*, Vol. 17, pp. 1-14.

[5] The size categories used for the Biotechnology Use and Development Survey 1999 are: Small: 50 or less employees; Medium: 51-150 employees; and Large: 151 and more employees.

[6] Some respondents reported more than 1 agreement.

[7] See this article for case studies on the nature of strategic alliances in the biotechnology sector.

Chapter Five

[1] The KNOW survey discussed here was funded by DG Research of the European Commission, under contract SOE2-CT98-1118. The authors would also like to thank A. Marcos Vera for his helpful comments on an earlier draft.

[2] See the collected research reports for the first CIS in Arundel and Garrelfs (1997), Baldwin and Da Pont (1996) for a similar survey in Canada, Francois and Favre (1998) for the second CIS for France, and Levin *et al.* (1987) for the Yale survey in the United States.

[3] An exception is a series of innovation surveys by Statistics Canada on biotechnology and on the use of Advanced Manufacturing Technologies, which ask an equal number of questions about internal and external knowledge sources.

[4] The percentage of Dutch firms in the five sectors surveyed by KNOW that innovate is available from Eurostat (1999) and is 62.7%, after weighting for the distribution of the samples drawn from each sector for the KNOW survey.

[5] It is not possible to use the results for the most important external knowledge source because cluster analysis assumes that every alternative is independent of the others.

[6] We also explored the effect of different knowledge sources on the economic value of product-based innovations to the firm, using the percentage of the firm's sales from the innovation. Only the coefficients for the firm's sector were significant.

[7] There is no correlation between the share of innovations developed in-house and the R&D employment share, indicating that these two variables measure different aspects of the firm's innovative capabilities.

[8] Small firms could also have a more limited range of internal capabilities, but this factor is captured by the variable for R&D employee share and the percent of innovations developed in-house. Therefore, the remaining effect of firm size is limited to the number of possible personal contacts.

[9] Including dummy variables for the most important idea source showed that only the use of suppliers for ideas had any effect on the regression. Firms that found suppliers to the most important external source of the original idea were less likely to find internal sources of greater importance than external sources. This suggests that, with this one exception, ideas have less of an influence on the value of external knowledge sources than those used for completion. This fits with other research that finds that it is less costly to obtain ideas than to develop ideas into commercially viable innovations. However, this result, while suggestive, may be due to the emphasis of the key question on the relative importance of internal and external knowledge sources on sources for completing the innovation.

Chapter Six

[1] This section draws on Jacques Lyrette, (1997), "Technological Entrepreneurship and Engineering in Canada", Canadian Academy of Engineering, Background Report, Chapter 8.

[2] The work of Craig S. Mullins of Platinum Technology Inc, titled, "What is Knowledge and can it be Managed", is helpful here and is available at: http://www.tdan.com/i008fe03.htm

[3] NSERC NCE annual report 1998-1999

[4] http://canarie.ca/about/quick_reference.html

[5] http://precarn.ca

[6] http://schoolnet.ca/home/e/whatis.asp

[7] Globe and Mail Jan 10, 2002.

[8] Innovation Management Association of Canada, "Managing Intellectual Capital", J. Lyrette NRC, Montreal, 2001

[9] Report of the NRC-NSERC Long Range Planning Panel, "The Origins of Structure in the Universe", 1999, August, p 44.

[10] Institute of Scientific Information Network http://www.isinet.com/isi/hot/research/200036/a

[11] AGRA-Coast now AMEC – Dynamic Structure Division

[12] Estimation of the Economic Impacts of Canadian Astronomy, The ARA Group, 2002 January.

[13] http :www.nrc.ca/IMI

[14] Statistics Canada, NRC contract 1985

[15] Cook, W.D. *et al.*, (1983). "Plant Level Productivity: A Comparison of Actual and Potential Performance", NRC contract.

[16] T. Calabrese and Joel Baum (U of T), and B. Silverman (Harvard): Canadian Biotechnology Start Ups, 1991-1997: The Role of Incumbents' Patents and Strategic Alliances in Controlling Competition., Social Sciences Research 29, 503-534 (2000). See:http://www.idealibrary.com and select the Social Science Research Journal.

[17] Unpublished research by Denys Cooper, National Research Council, 2001, updated data from item in The Conference Board of Canada's Third Innovation Report, (#236-01), November 2002, p 37 – 40 and Benedicte Callan, Analytical Report on High Tech Spin Offs (Paris; OECD, June 2000).

[18] Denys Cooper *ibid*

Chapter Seven

[1] Public Law 103-42, which amended the National Cooperative Research Act of 1984 (Public Law 98-462).

[2] In case of criminal or civil action, the collaborative activity of registered RJVs is evaluated under a rule of reason that determines whether the venture improves social welfare. Furthermore, any liability is limited to actual damages, rather than the usual treble damages under U.S. antitrust law (Link 1996; Vonortas 1997a).

[3] An analysis of trends in the figure is outside the scope of this chapter. See National Science Board (NSB) (2002).

[4] For the statutory definition of a research joint venture in the NCRA, see Link (1996).

[5] An analysis of trends in the figure is outside the scope of this chapter. See NSB (2002).

[6] Lehigh University, Bethlehem, PA. Web site: < http://www.lehigh.edu/~incbe/resea.htm >.

[7] For example, see NSB (2002), Brod and Link (2001), Vonortas (1997a), and Link (1996).

[8] See also Popper and Wagner (2001).

[9] For more on this topic see Feldman, *et al.*. (2001) and references therein.

[10] The latter focus on firm creation, not long term housing of firms.

[11] More recently, the Technology Transfer Commercialization Act of 2000 (Public Law 106-404) calls for certain changes in reporting requirements of Federal laboratories' technology transfer activities. These requirements are in the process of being implemented.

[12] Utterback (1994), Pavitt (1984), Vonortas (1997b), Sachwald (1998).

[13] A major exception is occupational data (including scientists, engineers, and related technical personnel) from the U.S. Bureau of Labor Statistics, which is based on an establishment survey, classified by industry.

Chapter Eight

[1] The author gratefully acknowledges the support of the Economic and Social Research Council's *Innovation Programme*, and the Design Council, in funding the original research referred to in this chapter.

[2] This case draws on Demaid (1998).

Chapter Nine

[1] This research has been supported, in part, by grants from the National Science Foundation and the Center for Innovation Management Studies at North Carolina State University. The authors thank the funding organizations for their support.

[2] A complex technology is one that cannot be understood in sufficient detail by an individual so that it can be communicated across time and distance with the precision necessary to be exactly reproduced. A simple technology can be understood.

[3] "Simple" and "complex" refer to the classifications of the process by which the product was made and/or of the product itself.

[4] Osborn and Hagedoorn (1997).

[5] Senker and Sharp (1997).

[6] See Rycroft and Kash (1999), pp.179-198, for a more detailed discussion of the three innovation patterns.

[7] Kash (2002a).

[8] Kash and Schaller (2000).

[9] Kash, Lieberman, and Schaller (2000).

[10] Kash (2002b).

[11] Kash (2001).

[12] Rycroft and Kash (1999), pp. 181-2.

[13] "Intel's Andy Grove, Inside," *Newsweek*, September 2, 1996, p. 63.

[14] Nelson and Winter (1982).

[15] Kline (1995), pp. 171-8.

[16] Sing (1997).

[17] The reader should note that the figure represents a general benchmark for comparison of the technologies under investigation. It does not provide precise metrics for the axes of decision-making and diversity of knowledge. In other words, the placement of innovation paths is not considered to indicate deviations in uniform degrees, rather it is intended to show general levels of difference.

[18] The selection of these variables was influenced in part by Hofstede (1980).

[19] Gannon (2001), pp. 157-176.

Chapter Ten

[1] "The general acceptance of the experimental method made it possible for hundreds even thousands of specialists to build the results of their individual research into a single store of information, usable across all sciences." (Rosenberg and Birdzell, 1986).

Chapter Eleven

[1] The remaining 5,037 patent applications come from other regions of the world.

[2] Other measures could well have been mobilised, namely the Gini index and the entropy measure of dispersion. All have in common their ability to measure some sort of concentration (Herfindhal and Gini index) or its inverse, namely the dispersion (Entropy). While the magnitude of their evolution might differ from one method to the other, all displayed similar patterns of evolution.

[3] Indeed, a careful examination of patents (see Appendix 1) shows a high concentration of patents in two central technological classes (genetic engineering and pharmaceuticals).

Chapter Twelve

[1] "Internet firms" are an ill-defined and heterogeneous group showing great volatility. There is no simple definition or classification of such firms. A description of larger US Internet-related firms can be found in OECD 2002a, Chapter 1, Appendix "ICT firms". This describes the top 25 US-based Internet-related firms listed on the New York Stock Exchange in October 2001 ranked by market capitalization. The activities of these firms varied very widely: six provided communication equipment and software; seven provided different kinds of on-line retail, reservation or trading services; six provided software for authentication, security and other e-business applications; and six provided ISP and portal services and related content services. These firms were spread out over a wide range of traditional US industrial classifications. Seven reported packaged software (US SIC 7372) as their primary business activity, two reported computer peripheral equipment manufacturing (SIC 3577), two reported computer systems design (7373), and two reported business services (SIC 7389). The remaining firms each reported different primary activities: book publishing and printing (US SIC 2731), telephone and telegraph apparatus manufacturing (3661), semiconductor manufacturing (3674), travel agency (4724), communications services (4899), security brokers and dealers (6211), hotels and motels (7011), computer programming services (7371), computer processing services

(7374), information retrieval services (7375), motion picture and video tape production (7812), and amusement and recreation services (7999).

In October 2001, the top 25 US-based Internet-related firms had a combined market capitalization of USD 331 billion. Just one of these firms (Cisco Systems) appeared in the top 50 IT firms ranked by total revenue in the most recent financial year ending prior to 30 September 2001. Three more (Broadcom, Juniper Networks and I2 Technologies) appeared in the top 250 IT firms ranked by revenue. During 2000-2001 they had combined revenues of USD 47 billion, more than USD 38 billion (82%) of which was earned by the top 10 firms. Revenue growth over the period 1998-2001 was rapid. 24 of these firms were already established in 1998 and increased their combined revenues by 247% from USD 13.5 billion in 1998 to USD 46.6 billion in 2001. These firms made a combined profit of USD 1.1 billion in 1998 falling to a combined loss of USD 28 billion in 2001. In 2001 Verisign Software contributed USD 15 billion to this combined loss, WebMD USD 4 billion, I2 Technologies USD 3 billion, and Amazon, Broadcom and Cisco Systems each contributed more than USD 1 billion, with 19 of the top 25 Internet firms showing losses in 2001.

[2] This work was initiated with a set of eleven case studies of firms engaged in business-to-business electronic commerce in France in three sectors: publishing, retail distribution and pharmaceuticals distribution [see OECD 2000a]. The approach adopted in this initial study was developed into a common methodology for the cross-country project. The TNO in the Netherlands, in conjunction with the Netherlands Ministry of Economic Affairs and the Telematica Instituut, worked with the OECD and with the French Ministry of Industry to develop the methodology and study plan. All of the work was in co-operation with the IPTS (part of the European Commission's Joint Research Centre). After extensive field-testing the methodology was applied in eleven countries to undertake internationally comparable firm-level case studies and use the case study information as the basis for preparing sector studies and national reports.

[3] The OECD statistical definitions for electronic commerce define the method by which the order is placed or received, not the payment or the channel of delivery, as determining whether the transaction is an Internet transaction (an order placed or received over the Internet) or an electronic transaction (an order placed or received over computer-mediated networks). In April 2001 the OECD developed guidelines for the interpretation of both Internet and computer-mediated network electronic commerce definitions. See OECD 2002d for a summary of this work.

[4] See OECD, 2002b. The sector reports and national reports are being progressively made public on the OECD web-site and are available at: http://www.oecd.org/oecd/pages/home/displaygeneral/0,3380,EN-document-29-nodirectorate-no-20-20570-29,FF.html

[5] See OECD 2002c, for a summary of the policy implications from this work.

[6] The EBIP work was reviewed at an OECD Workshop organized in Rome 29-30 October 2001 in conjunction with ISTAT (the Italian Statistical Institute) and the Italian government. Papers presented at the workshop are on the OECD web-site and are available at: http://www.oecd.org/EN/document/0, EN-document-29-nodirectorate-no-20-20435-29,00.html.

Chapter Thirteen

[1] I thank the organizers of the Workshop, particularly John de la Mothe, for the invitation to present an earlier version of this chapter, Fred Gault for his collegiality and insights, fellow participants in the Workshop for their suggestions and their spirit of lively exchange. The views expressed here are the author's own.

Chapter Fourteen

[1] Ralf Dahrendorf, "Old Wine in New Bottles", a December 2000 address before the London School of Economics and Political Science in the open society, p.1, communicated to this author as an attachment to a personal correspondence.

[2] See John de la Mothe and Gilles Paquet (eds.) (1999), *Information, Innovation and Impacts*, Kluwer, Boston; and John de la Mothe and Jorge Niosi, (2000), *The Economic and Social Dynamics of Biotechnology*, Kluwer, Boston.

[3] See John de la Mothe, "TechAction Town Halls", *Journal of Technology Transfer*, December, 2002 (in press)

[4] See, for an in depth summation, Richard Nelson and Sidney Winter, (1977), "Towards a Useful Theory of Innovation", *Research Policy*, 6, p.36-75; Richard Nelson and Sidney Winter, (1982), *Towards an Evolutionary Theory of Economic Change*, Belknap, Cambridge MA; and Giovanni Dosi *et al.*., (1988), *Economic Theory and Technical Change*, Pinter, London.

[5] John de la Mothe, (2000), "Government Science and the Public Interest", in G. Bruce Doern and Ted Reed (eds.), *Risky Business: Canada's Changing Science-Based Policy and Regulatory Regime*, University of Toronto Press, Toronto, p.31-48; and John de la Mothe, (1999), "Empowering Information and Networks Through Public Policy" in John de la Mothe and Gilles Paquet (eds.), *Information, Innovation and Impacts*, Kluwer, Boston, p.273-292

[6] Ronald S. Jonash and Tom Sommerlatte, (1999), *The Innovation Premium*, Perseus Books, New York; quoted on the IRAP-NRC web site.

[7] Quoted from the IRAP website homepage.

[8] Quoted from the IRAP website homepage.

BIBLIOGRAPHY

Abernathy, W. and Utterback, J. (1978). "Patterns of Industrial Innovation", *Technology Review*, Vol. 80, pp. 41-47.

Adams, J.D. (2001). "Comparative Localisation of Academic and Industrial Spillovers", NBER Working Paper Series 8292, Cambridge: National Bureau of Economic Research.

Afuah, A.N. and Utterback, J.M. (1997). "Responding to Structural Industry Change: A Technological Perspective", *Industrial and Corporate Change*, Vol. 6, No. 1, pp. 183-202.

Alchian, A.A. and Demsetz, H. (1972). "Production, Information Costs, and Economic Organization", *American Economic Review*, Vol. 62, No. 5, pp. 777-95.

Allen, T.J. (1977). *Managing the Flow of Technology*, Cambridge, MA: MIT Press.

Almeida, P., Song, J. and Grant, R.M. (2001). "Are Firms Superior to Alliances and Markets? An Empirical Test of Cross-border Knowledge Building", presented at Harvard Business School Strategy Research Conference, Oct 12-13, Cambridge, MA.

Alter, C. and Hage, J. (1993). *Organizations Working Together*, London: Sage.

Anderson, P. and Tushman, M.L. (1990). "Technological Discontinuities and Dominant Designs: A Cyclical Models of Technological Change", *Administrative Science Quarterly*, Vol. 35, No. 4, pp. 604-33.

Antonelli, C. (1999). "Industrial Organization and the Production of Knowledge", *Cambridge Journal of Economics*, Vol. 23, pp. 243-260.

Aoki, M. (1986). "Horizontal vs. Vertical Information Structure of the Firm", *American Economic Review*, Vol. 76, pp. 971-983.

Araujo, L. (1998). "Knowing and Learning as Networking", *Management Learning*, Vol. 29, No. 3, pp. 317-36.

Archibugi, D. (1992). "Patenting as an Indicator of Technological Innovation: A Review", *Science and Public Policy*, Vol. 19, No. 6, pp. 357-68.

Arora, A. and Gambardella, A. (1994). "The Changing Technology of Technological Change: General and Abstract Knowledge and the Division of Innovative Labor", *Research Policy*, Vol. 23, pp. 523-32.

Arrow, K.J. (1962). "Economic Welfare and the Allocation of Resources for Invention. The Rate and Direction of Inventive Activity", National Bureau of Economic Research, Princeton: Princeton University Press, pp. 609-625.

Arundel, A. and Garrelfs, R. (eds.) (1997). "Innovation Measurement and Policies", European Commission, EIMS publication 50.

Asanuma, B. (1985). "The Organization of Parts Purchases in the Japanese Automotive Industry", *Japanese Economic Studies*, pp. 32-53, Summer.

Asheim, B. and Isaksen, A. (2000). "Regional Innovation Policies Towards SMEs: Learning Good Practice from European Instruments", Workshop on the Regional Level of Implementation of Innovation, and Education and Training Policies, Brussels, November.

ASSENT (1999). "Working Towards Knowledge Society Telematics Applications - Programme Level Report: An Assessment of the Results of the Fourth Framework Telematics Application Programme", Report prepared for the European Commission DGXIII C/E. Brighton, The ASSENT (Assessment of Telematics) Project Consortium SU2101, principal responsibility of SPRU University of Sussex, Project SU 2101 - Deliverable D09.02. Part B (6), 30 June.

Audretsch, D.B., Link, A.N. and Scott, J.T. (2002). "Public/Private Partnerships: Evaluating SBIR-Supported Research", *Research Policy*, Vol. 31, pp. 145-158.

Auster, E.R. (1992). "The Relationship of Industry Evolution to Patterns of Technological Linkages, Joint Ventures, and Direct Investment between the U.S. and Japan", *Management Science*, Vol. 38, No. 6, pp. 778-792, June.

Barnett, W.-P. and Carroll, G.-R. (1987). "Competition and Mutualism among Early Telephone Companies", *Administrative Science Quarterly*, Vol. 32, pp. 400-21.

Baldwin, C. and Clark, K. (1997). "Managing in an Age of Modularity", *Harvard Business Review*, Sept-Oct.

Baldwin, J.R. (1997). "Importance of R&D for Innovation in Small and Large Canadian Manufacturing Firms", Statistics Canada, Catalogue No. 11F0019MPE, Analytical Studies Branch, Research Paper Series #107, Ottawa: Statistics Canada.

Baldwin, J.R. and Da Pont, M. (1996). "Innovation in Canadian Manufacturing Enterprises", Statistics Canada Catalogue no. 88-513-XPB, Ottawa: Industry Canada.

Baudeau, N. (1910). *Première introduction à la philosophie économique*, Paris: P. Geuthner.

Baum, J.A.C. and Oliver, C. (1991). "Institutional Linkages and Organizational Mortality", *Administrative Science Quarterly*, Vol. 36, pp. 187-98.

Baum, J.A.C. and Silverman, B. (1998). "Alliances, Patents, and Competitive Dynamics in Canadian Biotechnology" in Proceedings of the Annual Conference of the Administrative Sciences Association of Canada.

Baum, J.A.C., Calabrese, T. and Silverman, B.S. (2000). "Don't Go It Alone: Alliance Network Composition and Startups' Performance in Canadian Biotechnology", *Strategic Management Journal*, Vol. 21, pp. 263-94.

Bayona, C., Garcia-Marco, T. and Huerta, E. (2001). "Firms' Motivations for Cooperative R&D: an Empirical Analysis of Spanish Firms", *Research Policy*, Vol. 30, No. 8, pp. 1289-1308.

Bean, A.S., Ruse, M.J. and Whitely, R. (2000). "Benchmarking Your R&D: Results from IRI/CIMS Annual R&D Survey for FY '98", *Research Technology Management*, Vol. 43, No. 1, January-February.

Bell, D. (1973). *The Coming of Post-Industrial Society*, London: Heinemann.

Bernard, P. (1999). "La cohesion sociale: critique dialectique d'un quasi-concept", *Lien Social et Politiques*, Vol. 41, pp. 47-59.

Boland, R.J. Jr. and Tenkasi, R.V. (1995). "Perspective Making and Perspective Taking in Communities of Knowing", *Organization Science*, Vol. 6, No. 4, pp. 350-72.

Bordt, M., de Broucker, P., Read, C., Harris, S. and Zhang, Y. (2001a). "Science and Technology Skills: Participation and Performance in Elementary and Secondary School", *Education Quarterly Review*, Vol. 8, No. 1, pp. 12-21.

Bordt, M., de Broucker, P., Read, C., Harris, S. and Zhang, Y. (2001b). "Science and Technology Skills: Participation and Performance in University and Beyond", *Education Quarterly Review*, Vol. 8, No. 1, pp. 22-34.

Bordt, M., de Broucker, P., Read, C., Harris, S. and Zhang, Y. (2001c). "Determinants of Science and Technology Skills: Overview of the Study", *Education Quarterly Review*, Vol. 8, No. 1, pp. 8-11.

Breshi, S., Malerba, F. and Orsenigo, L. (2000). "Technological Regimes and Schumpeterian Patterns of Innovation", *The Economic Journal*, Vol. 110, pp. 388-410, April.

Bressand, A. and Kalypso, N. (eds) (1989). *Strategic Trends in Services: An Inquiry into the Global Service Economy*, New York: Harper and Row.

Brod, A.C. and Link, A. (2001). "Trends in Cooperative Research Activity", in Feldman, M.P. and Link, A. (eds.), *Technology Policy for the Knowledge-based Economy*, Boston: Kluwer.

Brooks, H. (1994). "The Relationship between Science and Technology", *Research Policy*, Vol. 23, pp. 477-86.

Brown, J.S. and Duguid, P. (1998). "Organizing Knowledge", *California Management Review*, Vol. 40, No. 3.

Brown, P. and Lauder, H. (2001). *Capitalism and Social Progress: The Future of Society in a Global Economy*, Houndsmills, Basingstroke, UK: Palgrove.

Burt, R. (1976). "Positions in Social Networks", *Social Forces*, Vol. 55, pp. 93-122.

Burt, R. (1992). *Structural Holes: The Social Structure of Competition*, Cambridge, MA: Harvard University Press.

Burt, R. (1997). "The Contingent Value of Social Capital", *Administrative Science Quarterly*, Vol. 42, pp. 339-65.

Callon, M. (1994). "Is Science a Public Good?", *Science, Technology and Human Values*, Vol. 19, No. 4, pp. 395-425.

Callon, M. and Law, J. (1997). "After the Individual in Society: Lessons on Collectivity from Science and Technology", *The Canadian Journal of Sociology*, Vol. 22, No. 2, pp. 165-182.

Carlsson, B. (2000). "A Systems Framework for the Study of Economic and Social Dynamics of Biotechnology", in de la Mothe, J. and Niosi, J. (eds.), *The Economic and Social Dynamics of Biotechnology*, Boston: Kluwer.

Carr, R.K. (1995). "U.S. Federal Laboratories and Technology Transfer", SRI International, Mimeograph.

Cassiman, B., Pérez-Castrillo, D. and Veugelers, R. (2002). "Endogeneizing Know-How Flows through the Nature of R&D Investments", *International Journal of Industrial Organization*, Volume 20, No. 6. pp. 775-799.

Castells, M. (1999). "The Social Implications of Information and Communication Technologies", in *The World Science Report*, Paris: UNESCO.

Chandler, A.D. (1977). *The Visible Hand: The Managerial Revolution in American Business*, Cambridge: Belknap.

Chesbrough, H. and Teece, D.J. (1996). "When is Virtual Virtuous: Organizing for Innovation", *Harvard Business Review*, pp. 65-74, January-February.

Chesbrough, H.W. and Kusunoki, K. (2001). "The Modularity Trap: Innovation, Technology Phase-Shifts and the Resulting Limits of Virtual Organizations", in Teece, D. (ed), *Managing Industrial Knowledge*, New York: Russell Sage Press.

Coburn, C. (1995). *Partnerships: A Compendium of State and Federal Technology Programs*, Columbus, OH: Battelle Press.

Cohen, W.M. and Levinthal, D. (1980). "Innovation and Learning: the Two Faces of R&D". *The Economic Journal*, Vol. 99, pp. 569-596.

Cohen, W.M. and Levinthal, D.A. (1989). "Innovation and Learning: Two Faces of R&D", *Economic Journal*, Vol. 99, pp. 569-596.

Cohen, W.M. and Levin, R.C. (1989). "Empirical Studies of Innovation and Market Structure", in Schmalensee, R. and Willig, R.D. (eds.), *Handbook of Industrial Organization*, Amsterdam: North-Holland, pp. 1059-1107.

Cohen, W.M. and Levinthal, D.A. (1990). "Absorptive Capacity: A New Perspective on Learning and Innovation", *Administrative Science Quarterly*, Vol. 35, pp. 128-152.

Cohen, W.M., Florida, R. and Goe, W.R. (1994). *University-Industry Research Centers*, Pittsburgh: Carnegie Mellon University.

Cohen, W.M., Florida, R., Randazzese, L. and Walsh, J. (1998). "Industry and the Academy: Uneasy Partners in the Cause of Technological Advance", in Noll, R. (ed.), *Challenge to the University*, Washington, DC: Brookings Institution Press, pp. 171-199.

Coleman, J. (1990). *Foundations of Social Theory*, Cambridge, MA: Harvard University Press.

Coleman, W.D. and Perl, A. (1997). "Internationalized Policy Environments and Policy Network Analysis", paper presented to the Annual Meeting of the American Political Science Association, Washington, D.C.

Contractor, F.J. and Lorange, P. (1988). "Why Should Firms Cooperate? The Strategy and Economics Basis for Cooperative Ventures", in Contractor, F.J. and Lorange, P. (eds.), *Cooperative Strategies in International* Business, Lexington: Lexington Books, pp. 3-30.

Cook, S.D.N. and Brown, J.S. (1999). "Bridging Epistemologies: The Generative Dance Between Organizational Knowledge and Organizational Knowing", *Organization Science*, Vol. 10, No. 4, pp. 381-400.

Coombs, R., Harvey, M. and Tether, B. (2001). "Analysing Distributed Innovation Processes", CRIC Discussion Paper 43, Manchester: University of Manchester.

Copper, R.S. and Merrill, S. (1998). "Trends in U.S. Industrial Innovation: An Assessment of National Data Sources and Information Gaps", Science, Technology, and Economic Policy Board (STEP) of the U.S. National Research Council (NRC), Washington: NRC.

Curtis, B. (2001). *The Politics of Population: State Formation, Statistics, and the Census of Canada 1840-1875*, Toronto: University of Toronto Press.

Dahrendorf, R. (2000). "Old Wine in New Bottles", address before the London School of Economics and Political Science in the open society, December.

Dasgupta, P. and David, P. (1994). "Toward a New Economics of Science", *Research Policy*, Vol. 23, No. 5, pp. 487-522.

David, P.A., Mowery, D. and Steinmuller, E. (1994). "University-Laboratory Research Collaborations: Managing Missions in Conflict", University Goals, Institutional Mechanisms and the Industrial Transferability of Research, Stanford: Stanford University.

David, P.A. and Foray, D. (1994). "Percolation Structures, Markov Random Fields and the Economics of EDI Standards Diffusion", in Pogorel, G. (ed.), *Global Telecommunication Strategies and Technological Changes*, Elsevier Science.

Davidow, W.H. and Malone, M.S. (1992). *The Virtual Corporation: Structuring and Revitalizing the Corporation for the 21st Century*, New York: Harper Business.

de la Mothe, J. (1994). "Canada and the National System of Innovation'', in Resource Book for Science and Technology Consultation, Government of Canada, Vol II, pp.11-20.

de la Mothe, J. (1999). "Empowering Information and Networks through Public Policy", in de la Mothe, J. and Paquet, G. (eds.), *Information, Innovation and Impacts*, Boston: Kluwer, pp.273-292.

de la Mothe, J. (2000). "Government Science and the Public Interest", in Doern, G.B. and Reed, T. (eds.), *Risky Business: Canada's Changing Science-Based Policy and Regulatory Regime*, Toronto: University of Toronto Press, pp.31-48.

de la Mothe, J. (2002). "TechAction Town Halls", *Journal of Technology Transfer*, December (in press).

de la Mothe, J. and Paquet, G. (eds.) (1996). *Evolutionary Economics and the New International Political Economy*, London: Pinter.

de la Mothe, J. and Paquet, G. (eds.) (1998a). *Local and Regional Systems of Innovation*, Boston: Kluwer.

de la Mothe, J. and Paquet, G. (1998b). "National Systems of Innovation, 'Real Economies', and Instituted Proesses", *Small Business Economics*, Vol. 11, pp.101-111.

de la Mothe, J. and Paquet, G. (eds.) (1999). *Information, Innovation and Impacts*, Boston: Kluwer.

de la Mothe, J. and Niosi, J. (eds.) (2000). *The Economic and Social Dynamics of Biotechnology*, Boston: Kluwer.

de la Mothe, J. and Foray, D. (eds.) (2001). *Knowledge Management in the Innovation Process*, Boston: Kluwer.

DeBresson, C. and Amesse, F. (1991). "Networks of Innovators: A Review and Introduction to the Issue", *Research Policy*, Vol. 20, pp. 363-379.

Demaid, A. (1998). "Managing Knowledge Across the Cultural Divides", paper presented at the Management of Knowledge and Innovation Research Unit Workshop, 30 September, London.

Dodgson, M. (1991). *The Management of Technological Learning: Lessons from a Biotechnology Company*, New York: Walter de Gruyter.

Dodgson, M. and Rothwell, R. (eds.) (1994). *The Handbook of Industrial Innovation*. Aldershot: Edward Elgar.

Dosi, G. (1982). "Technological Paradigms and Technological Trajectories: A Suggested Interpretation of the Determinants and Directions of Technical Change", *Research Policy*, Vol. 11, pp. 147-62.

Dosi, G. *et al.*., (1988). *Economic Theory and Technical Change*, London: Pinter.

Drucker, P.F. (1969). *The Age of Discontinuity: Guidelines to Our Changing Society*, London: Heinemann.

Dufour, P.R. and de la Mothe, J. (eds.) (1993). *Science and Technology in Canada*, London: Longman.

Dyer, J.H. and Singh, H. (1998). "The Relational View: Cooperative Strategy and Sources of Interorganizational Competitive Advantage", *Academy of Management Review*, Vol. 23, pp. 660-79.

Earl, L. (2002). "Innovation and Change in the Public Sector: A Seeming Oxymoron", Science, Innovation and Electronic Information Division, Working Papers, Catalogue number 88F0006XIB, no. 1., Ottawa: Statistics Canada, (www.statcan.ca/english/research/88F0006X1B/88F0006XIE2002001.pdf)

Edquist, C. (ed.) (1997). *Systems of Innovation: Technologies, Institutions and Organizations*, London: Pinter.

Eliasson, G. (2000). "The Industrial Potential of Biotechnology: A Competence Bloc Analysis", in de la Mothe, J. and Niosi, J. (eds.), The *Economic and Social Dynamics of Biotechnology*, Boston: Kluwer.

Ernst&Young (1998). "New Directions 98: The Twelfth Biotechnology Industry Annual Report", 49.

Ernst&Young (2000). "Evolution: Ernst & Young's Seventh Annual European Life Sciences Report 2000", 71.

Ernst&Young (2001). "Integration: Ernst & Young's Eighth Annual European Life Sciences Report 2001", 79.

Eurostat (1999). "Statistics In Focus, Research and Development, Theme 9 – 2/1999. Community Innovation Survey 1997/98," Eurostat.

Evans, R.G. and Stoddart, G.L. (1994). "Producing Health, Consuming Health Care", in Evans, R.G., Barer, M.L. and Marmor, T.R. (eds.), *Why Are Some People Healthy and Others Not? The Determinants of Health of Populations*, New York: Aldine de Gruyter, pp. 27-64.

Fagan, D. (2002). "Innovation Diagnosis is in; Now for the Remedy", *The Globe and Mail*, 13 February, B10.

Faulkner, W., Senker, J. and Velho, L. (1995). *Knowledge Frontiers: Public Sector Research and Industrial Innovation in Biotechnology, Engineering Ceramics, and Parallel Computing*, Oxford: Clarendon Press.

Feldman, M.P. (2001). "Strategic Research Partnerships in Biotechnology" in Strategic Research Partnerships: Proceedings from an NSF Workshop, Arlington VA.

Feldman, M.P., Feller, I., Bercovitz, J.E.L. and Burton, R.M. (2001). "Understanding Evolving University-Industry Relationships", in Feldman, M.P. and Link, A. (eds.), *Technology Policy for the Knowledge-Based Economy*, Boston: Kluwer.

Foucault, M. (1991). "Governmentality", in Burchell, G., Gordon, C. and Miller, P. (eds.), *The Foucault Effect: Ethics, Studies in Governmentality*, Chicago: University of Chicago Press, pp. 87-104.

Francois, J.-P. and Favre, F. (1998). "Technological Innovation is Progressing in Industry", SESSI: Le 4 Pages des Statistiques Industrielles, Paris: SESSI.

Freeman, C. (1967). "Science and the Economy at the National Level", in OECD, *Problems in Science Policy*, Paris: OECD.

Freeman, C. (1978). *The Economics of Industrial Innovation*, London: Penquin, Harmonsworth.

Freeman, C. (1982). *The Economics of Industrial Innovation*, London: Pinter.

Freeman, C. (1991). "Networks of Innovators: A Synthesis of Research Ideas", *Research Policy*, Vol. 20, pp. 499-514.

Freeman, C. (1992). *The Economics of Hope*, London: Pinter.

Freeman, C. and Louçã, F. (2001). *As Time Goes By: From the Industrial Revolutions to the Information Revolution*, Oxford: Oxford University Press.

Freeman, L. (1979). "Centrality in Social Networks: Conceptual Clarification", *Social Networks*, Vol. 1, pp. 215-39.

Frenken, K. (1998). "A Complexity Approach to Technological Evolution," Conférence ETIC. Strasbourg, 1-15.

Galt, V. (2002). "Medicare Cut Seen Raising Labor Costs", *The Globe and Mail*, 12 February, B10.

Gannon, M.J. (2001). *Understanding Global Cultures: Metaphorical Journeys Through 23 Nations*, Thousand Oaks: Sage Publications.

Gelsing, L. (1989). "Knowledge Networks, Industrial Flexibility and Innovation", in Borum, F. and Kristensen, P. (eds.), *Technological Change and Organizational Change*, Copenhagen: Danish Social Sciences Research Council.

Gelsing, L. (1992). "Innovation and the Development of Industrial Networks', in Lundvall, B.-A. (ed.), *National Systems of Innovation: Towards a Theory of Innovation and Interactive Learning*, London: Pinter, pp. 116–27.

Georghiou, L. (1998). "Science, Technology and Innovation Policy for the 21st Century", *Science and Public Policy*, Vol. 25, pp.135-139.

Gibbons, M., Limoges, C., Nowotny, H., Schwartzman, S., Scott, P. and Trow, M. (1994). *The New Production of Knowledge: The Dynamics of Science and Research in Contemporary Societies*, London: Sage Publications.

Granstrand, O., Patel, P. and Pavitt, K. (1997). "Multi-Technology Corporations: Why they have 'Distributed' rather than 'Distinctive Core' Competencies", *California Management Review*, Vol. 39, No. 4, pp. 8-25.

Grant, R.M. (1996). "Towards a Knowledge-Based Theory of the Firm", *Strategic Management Journal*, Vol. 17, (Special Winter Issue), pp. 109-122.

Greis, N.P., Dibner, M.D. and Bean, A.S. (1995). "External Partnering as a Response to Innovation Barriers and Global Competition in Biotechnology", *Research Policy*, Vol. 24, No.4, pp. 609-630.

Griliches, Z. (1958). "Research Costs and Social Returns: Hybrid Corn and Related Innovations," *Journal of Political Economy*, Vol. 66, pp. 501-522.

Griliches, Z. (1979). "Issues in Assessing the Contribution of Research and Development to Productivity Growth", *Bell Journal of Economics*, Vol. 10, No. 1, pp. 92-116.

Griliches, Z. (1990). "Patents Statistics as Economic Indicators: A Survey", *Journal of Economic Literature*, Vol. 28, pp. 1661-707.

Haas, P.M. (1992). "Introduction: Epistemic Communities and International Policy Coordination", *International Organization*, Vol. 46, No. 1, Winter.

Hagedoorn, J. (1990). "Organizational Modes of Inter-Firm Co-Operation and Technology Transfer", *Technovation*, Vol. 10, No. 1, pp. 17-30.

Hagedoorn, J. (1993). "The Rationale of Strategic Technology Partnering: Interorganizational Modes of Cooperation and Sectoral Differences", *Strategic Management Journal*, Vol. 14, pp. 371-385.

Hagedoorn, J. and Schakenraad, J. (1990). "Interfirm Partnerships and Cooperative Strategies in Core Technologies", in Freeman, C. and Soete, L. (eds.), *New Explorations in the Economics of Technical Change*. London: Pinter.

Hagedoorn, J. and Schakenraad, J. (1992). "Leading Companies and Networks of Strategic Alliances in Information Technologies", *Research Policy*, Vol. 21, pp. 163-90.

Hagedoorn, J., Link, A.N. and Vonortas, N.S. (2000). "Research Partnerships", *Research Policy*, Vol. 29, pp. 567-586.

Haggerty, K.D. (2001). *Making Crime Count*, Toronto: University of Toronto Press.

Haklisch, C.S. and Pouletty, P. (1986). "Technical Alliances in the Semiconductor Industry", Center for Science and Technology Policy, New York University, New York.

Hamel, G., Doz, Y.L. and Prahalad, C.K. (1989). "Collaborate with Your Competitors – and Win", *Harvard Business Review*, pp. 133-139, January-February.

Hayek, F. (1937). "Economics and Knowledge", *Economica*, Vol. 4, pp. 33-54.

Hébert, R.F. and Link, A.N. (1988). *The Entrepreneur: Mainstream Views and Radical Critiques*, New York: Praeger Academic Publishers.

Helliwell, J. (2001). "Social Capital, the Economy and Well-Being", in Banting, K., Sharpe, A. and St.-Hilaire, F. (eds.), *The Review of Economic Performance and Social Progress, The Longest Decade: Canada in the 1990s*, Montreal and Ottawa: The Institute for Research on Public Policy and Centre for the Study of Living Standards.

Hennart, J.-F. (1988). "A Transaction Cost Theory of Equity Joint Ventures", *Strategic Management Journal*, Vol. 9, No. 4, pp. 361-374.

Hertzfeld, H.R., Link, A.N. and Vonortas, N.S. (2001). "Intellectual Property Protection Mechanisms and Research Partnerships", paper presented at the Workshop on Intellectual Property Protection Mechanisms in Research Partnerships, Center for International Science and Technology Policy, George Washington University, Washington, D.C., October 2001.

Hicks, D. and Katz, S. (1997). "Published Papers, Tacit Competencies and Corporate Management of the Public/Private Character of Knowledge", *Industrial and Corporate Change*, Vol. 4, pp. 401-24.

Hippel, E. von (1988). *The Sources of Innovation*, Oxford: Oxford University Press.

Hippel, E. von (1990). "The Impact of Sticky Data on Innovation and Problem Solving", Sloan School of Management Working Paper #3147-90-BPS, April.

Hippel, E. von (1994). "Sticky Information and the Locus of Problem Solving: Implications for Innovation", *Management Science*, Vol. 40, No. 4, pp. 429-439.

Hite, J.M. and Hesterly, W.S. (2001). "Research Notes and Commentaries. The Evolution of Firm Networks: From Emergence to Early Growth of the Firm", *Strategic Management Journal*, Vol. 22, pp. 275-86.

Hofstede, G. (1980). *Culture's Consequences: International Differences in Work-Related Values,* (Abridged Edition), Newbury Park, CA: Sage Publications.

Howells, J. (1999). "Research and Technology Outsourcing", *Technology Analysis and Strategic Management*, Vol. 11, pp. 17-29.

Human Resources Development Canada (HRDC) (2002). "Achieving Excellence: Investing in People, Knowledge and Opportunities", Ottawa: Industry Canada, (www.innovationstrategy.gc.ca)

Industry Canada (2002). "Knowledge Matters: Skills and Learning for Canadians", Ottawa: Industry Canada, (www.innovationstrategy.gc.ca)

Jankowski, J.E., Link, A.N. and Vonortas, N.S. (2001). "Strategic Research Partnerships: Proceedings from an NSF Workshop", Arlington, VA: National Science Foundation.

Jervis, R. (1997). *System Effects: Complexity in Political and Social Life*, Princeton: Princeton University Press

Jewkes, J., Sawers, D., Stillerman, R. (1969). *The Sources of Innovation*, London: Macmillan.

Jonash, R.S. and Sommerlatte, T. (1999). *The Innovation Premium*, New York: Perseus Books.

Kamien, M.I. and Zang, I. (2000). "Meet Me Halfway: Research Joint Ventures and Absorptive Capacity", *International Journal of Industrial Organization*, Vol. 18, pp. 995-1012.

Kash, D.E. (2001). "Haier Company Innovation in Household Appliances", School of Public Policy Working Paper, (November), Fairfax, VA: George Mason University.

Kash, D.E. (2002a). "Bosch Fuel Injection Systems for Diesel Engines", School of Public Policy Working Paper, (February), Fairfax, VA: George Mason University.

Kash, D.E. (2002b). "TATA Consultancy Services: A Case Study", School of Public Policy Working Paper, (February), Fairfax, VA: George Mason University.

Kash, D.E. and Rycroft, R.W. (1994). "Technology Policy: Fitting Concept with Reality", *Technological Forecasting and Social Change*, Vol. 47, pp. 35-48.

Kash, D.E. and Schaller, R.R. (2000). "Innovation of the Audio Compact Disc Player", The Institute of Public Policy Working Paper, (January), Fairfax, VA: George Mason University.

Kash, D.E., Lieberman, S.L. and Schaller, R.R. (2000). "Innovation of Cardio-Imaging Technology at Hewlett-Packard and HP/Philips", The Institute of Public Policy Working Paper, (January), Fairfax, VA: George Mason University.

Katz, J.S. and Martin, B.R. (1997). "What is Research Collaboration?", *Research Policy*, Vol. 26, pp. 1-18.

Kelley, M. and Arora, A. (1996). "The Role of Institution-Building in US Industrial Modernisation Programs", *Research Policy,* Vol. 25.

Kenney, M. (1986). "Biotechnology: The University - Industrial Complex", New Haven: Yale University Press.

Kernaghan, K. and Siegel, D. (2001). *Public Administration in Canada,* (4th Edition), Toronto: Nelson.

Klepper, S. (1997). "Industry Life Cycles", *Industrial and Corporate Change*, Vol. 6, No. 1, pp. 145-201.

Kline, S.J. (1995). *Conceptual Foundations for Multidisciplinary Thinking*, Stanford: Stanford University Press.

Kodama, F. (1995). *Emerging Patterns of Innovation: Sources of Japan's Technological Edge*, Boston, MA: Harvard Business School Press.

Kogut, B. (1988). "Joint Ventures: Theoretical and Empirical Perspectives", *Strategic Management Journal*, Vol. 9.

Kogut, B. (2000). "The Network as Knowledge: Generative Rules and the Emergence of Structure", *Strategic Management Journal*, Vol. 21, No. 3, pp. 405-25.

Kogut, B. and Zander, U. (1996). "What Do Firms Do? Coordination, Identity, and Learning", *Organization Science*, Vol. 7, No. 5, pp. 502-518.

Kondro, W. (2002). "Innovation White Papers", *Perspectives: An Electronic Newsletter on Research and Science Policy*, Vol. 5, No. 4, 18 February, (www.hssfc.ca/english/policyand advocacy/perspectives)

KPMG (2002). "Comparing Business Costs in North America, Europe and Japan", (www.competitivealternatives.com)

Kuhn, T. (1962). *The Structure of Scientific Revolutions*, Chicago: Chicago University Press.

Lacroix, J.-G. and Tremblay, G. (1997). "The 'Information Society' and Cultural Industries Theory", *Current Sociology*, Vol. 45, No. 4.

Landes, D.S. (1969). *The Unbound Prometheus: Technological Change 1750 to the Present*, Cambridge: Cambridge University Press.

Langlois, R.N. (1992). "Transaction-Cost Economics in Real Time", *Industrial and Corporate Change*, Vol. 1, pp. 99-127.

Langlois, R.N. and Robertson, P. (1989). "Explaining Vertical Integration: Lessons from the American Automobile Industry", *The Journal of Economic History*, Vol. 49, No. 2.

Langlois, R.N. and Foss, N.J. (1999). "Capabilities and Governance: The Rebirth of Production in the Theory of Economic Organization", *Kyklos*, Vol. 52, pp. 201-218.

Langrish, J., *et al.* (1972). *Wealth from Knowledge: A Study of Innovation in Industry*, New York: Halsted Press Division: John Wiley and Sons.

Latour, B. (1987). *Science in Action: How to Follow Scientists and Engineers Through Society*, Cambridge, MA: Harvard University Press.

Latour, B. (1988). *The Pasteurization of France*, (translated by Sheridan, A. and Law, J.), Cambridge, MA: Harvard University Press.

Lave, J. (1993). "The Practice of Learning', in Chaiklin, S. and Lave, J. (eds.), *Understanding Practice. Perspectives on Activity and Context*, Cambridge: Cambridge University Press.

Lave, J. and Wenger, E. (1991). *Situated Learning: Legitimate Peripheral Participation*, Cambridge: Cambridge University Press.

Leonard, D. (1995). *Wellsprings of Knowledge: Building and Sustaining the Sources of Innovation*, Boston, MA: Harvard Business School Press.

Lerner, J. and Merges, R. (1997). *The Control of Strategic Alliances: An Empirical Analysis of Biotechnology Collaborations*, Cambridge, MA: National Bureau of Economic Research.

Lerner, J. and Merges, R. (1998) "The Control of Technology Alliances: An Empirical Analysis of the Biotechnology Industry", *The Journal of Industrial Economics*, Vol. 45, No. 2, pp. 125-156.

Levin, R.C. *et al.* (1987). "Appropriating the Returns from Industrial Research and Development", Brookings Papers on Economic Activity, Vol. 3, pp. 242-279.

Levina, N. (2001). "Sharing Knowledge in Heterogeneous Environments, Reflections", *The SoL Journal*, Vol. 2, No. 2.

Liebeskind, J.P., Oliver, A.L., Zucker, L. and Brewer, M. (1996). "Social Networks, Learning and Flexibility: Sourcing Scientific Knowledge in Biotechnology Firms", *Organization Science*, Vol. 7, No. 4, pp. 428-43.

Link, A.N. (1987). *Technological Change and Productivity Growth*, London: Harwood Academic Publishers.

Link, A.N. (1996). "Research Joint Ventures: Patterns from Federal Register Filings", *Review of Industrial Organization*, pp. 617-628, October.

Link, A.N. (1999). "Public/Private Partnerships in the United States", *Industry and Innovation*, Vol. 6, pp. 191-217.

Link, A.N. and Tassey, G. (1987). *Strategies for Technology-based Competition: Meeting the New Global Challenge*, Lexington, MA: Lexington Books.

Link, A.N., Teece, D.J. and Finan, W.F. (1996). "Estimating the Benefits from Collaboration: The Case of SEMATECH", *Review of Industrial Organization*, Vol. 11, pp. 737-751.

Link, A.N. and Scott, J.T. (1998). *Public Accountability: Evaluating Technology-Based Institutions*, Norwell, MA: Kluwer.

Link, A.N. and Scott, J.T. (2000). "Explaining Observed Licensing Arrangements: Towards a Broader Understanding of Technology Flows", Dartmouth College Working Paper 00-09.

Link, A.N. and Scott, J.T. (2001). "The Growth of U.S. Science Parks", paper presented at the Allied Social Sciences Association Meeting, New Orleans, January.

Link, A.N. and Scott, J.T. (2002). "Science Parks and the Academic Mission of Universities: An Exploratory Study", paper presented at the Georgia Institute of Technology Roundtable for Engineering Entrepreneurship Research Conference, Atlanta, Georgia, March 2002.

Little, B. (2002). "Evidence Points to No Recession in Canada", *The Globe and Mail*, 23 February, B7.

Loasby, B.J. (1996). "The Division of Labor", *History of Economic Ideas*, Vol. 4, No. 1-2, pp. 299-323.

Loasby, B.J. (1999). *Knowledge, Institutions and Evolution in Economics*, London: Routledge.

Loasby, B.J. (2000a). "Market Institutions and Economic Evolution", *Journal of Evolutionary Economics*, Vol. 10, No. 3, pp. 297-309.

Loasby, B.J. (2000b). "The Division and Organization of Knowledge", *European Journal of Economic and Social Systems*, Vol. 14, No. 2, pp. 143-55.

Loasby, B.J. (2001). "Industrial Dynamics: Why Connections Matter", DRUID Academy Winter Conference, Klarkovgaard, January 18-20.

Lundvall, B.-A., (ed.) (1992). *National Systems of Innovation: Towards a Theory of Innovation and Interactive Learning*, London: Pinter.

Lunzer, F. (1988). "Cash Crisis Creates Biotech Alliances", *High Technology Business*, April, pp. 18-23.

Machlup, F. (1962). *The Production and Distribution of Knowledge in the United States*, New Jersey: Princeton University Press.

Malerba, F. (1992). "Learning by Firms and Incremental Technical Change", *The Economic Journal*, Vol. 102, pp. 845-59.

Malerba, F. and Orsenigo, L. (1997). "Technological Regimes and Sectoral Patterns of Innovative Activities", *Industrial and Corporate Change*, Vol. 6, No. 1, pp. 83-117.

Mansfield, E. (1991). "Academic Research and Industrial Innovation", *Research Policy*, Vol. 20, pp. 1-12.

Mansfield, E. (1995). "Academic Research Underlying Industrial Innovations: Sources, Characteristics and Financing", *Review of Economics and Statistics*, Vol. 75, No. 1, pp. 55-65.

Mansfield, E., Rapoport, J., Romeo, A., Wagner, S. and Beardsley, G. (1977). "Social and Private Rates of Return from Industrial Innovations", *Quarterly Journal of Economics*, Vol. 91, pp. 221-240.

March, J.G. (1991a). "Exploration and Exploitation in Organizational Learning", *Organization Science*, Vol. 2, No. 1.

Martin, S. and Scott, J.T. (2000). "The Nature of Innovation Market Failure and the Design of Public Support for Private Innovation", *Research Policy*, Vol. 29, pp. 437-447.

McDaniel, S.A. (2000). "Capturing the Elusive Social Impacts of Science and Technology: Towards a Research Agenda", in de la Mothe, J. and Paquet, G. (eds.), *Information, Innovation and Impacts*, Boston: Kluwer, pp. 109-124.

McDaniel, S.A. (2002). "Information and Communications Technologies: Bugs in the Generational Ointment?", plenary paper invited for 2001 BUGS: International Sociological Association Conference, Montreal, April 2002.

McKelvey, M. (1996). "Defining Transfer in Biotechnology and Software: Multiple Creation of Knowledge and Issues of Ownership", in, Inzelt, A. and Coenen, R. (eds.) , *Knowledge, Technology Transfer and Foresight*, Boston: Kluwer.

McNamara, P. and Baden-Fuller, C. (1999). "Lessons From the Celltech Case: Balancing Knowledge Exploration and Exploitation in Organizational Renewal", *British Journal of Management*, Vol. 10, pp. 291-307.

McNiven, C. (2001). "Biotechnology Use and Development – 1999", SIEID Working Paper #7, Catalogue 88F0006XIE-No. 7, Ottawa: Statistics Canada.

Meisel, J. (1998). "Caesar and the Savants: Some Socio-Political Contexts of Science and Technology in Canada", in Herzberg, A.M. and Krupka, I. (eds.), *Statistics, Science and Public Policy*, Kingston, ON: Queen's University, pp. 153-177.

Melkers, J. (1993). "Bibliometrics as a Tool for Analysis of R&D Impacts", in Bozeman, B. and Melkers, J. (eds.), *Evaluating R&D Impacts: Methods and Practice*, Boston: Kluwer.

Merton, R. (1973). *The Sociology of Science: Theoretical and Empirical Investigations*, Chicago: Chicago University Press.

Miles, R.E. and Snow, C.C. (1994). *Fit, Failure, and the Hall of Fame: How Companies Succeed or Fail*, Toronto: Maxwell MacMillan Canada.

Mowery, D.C. (1995). In Stoneman, P. (1995). *Handbook of the Economics of Innovation and Technological Change*, Oxford: Blackwell, pp. 513-557.

Murray, F. (2001). "Markets for Scientific Ideas: An Exploration of the Commercialization of Biomedical Science", R. Calori, EGOS 17th annual conference. Lyon (FRA):, 2001.

Nantua, R., Michel, D. and Salle, R. (1999). "Du solo à la cordée : réussir le lancement d'une innovation par l'établissement d'un standard professionnel compatible – le cas ASB° chez SNR Roulements", *Décisions Marketing*, No. 17, May-August.

National Science Board (NSB) (2002). "U.S. and International Research and Development: Funds and Alliances", in Science and Engineering Indicators - 2002, NSB-02-1. Arlington, VA: National Science Foundation.

Nelson, R.R. (1959). "The Simple Economics of Basic Scientific Research", *Journal of Political Economy*, pp. 297-306, June.

Nelson, R.R. (ed.) (1993). *National Systems of Innovation*, Oxford: Oxford University Press.

Nelson, R.R. and Winter, S.G. (1977), "Towards a Useful Theory of Innovation", *Research Policy*, Vol. 6, pp.36-75

Nelson, R.R. and Winter, S.G. (1982). *An Evolutionary Theory of Economic Change*, Cambridge, MA: Belknap Press of Harvard University Press.

Nesta, L. and Saviotti, P.P. (2002). "The Coherence of the Knowledge Base and the Firm's Innovative Performance. Evidence from the Us Bio-Pharmaceutical Industry", forthcoming, pp. 30 .

Newsweek (1996). "Intel's Andy Grove, Inside," September 2, 1996, p. 63.

Niosi, J. (1996). "Strategic Technological Collaboration in Canadian Industry: Towards a Theory of Flexible or Collective Innovation", in Coombs, R. *et al.* (eds.), *Technological Collaboration: The Dynamics of Cooperation in Industrial Innovation*, Brookfield, VT: Edward Elgar.

Niosi, J. (2000). *Explaining Rapid Growth In Canadian Biotechnology Firms*, Statistics Canada, Cat. No. 88F0017MIE, SIEID Research Paper # 8, Ottawa: Statistics Canada.

Nonaka, I. and Takeuchi, H. (1995). *The Knowledge-Creating Company*. Oxford: Oxford University Press.

OECD/Eurostat (1997). *Proposed Guidelines for Collecting and Interpreting Technological Innovation Data, 'Oslo Manual'*, Paris: OECD.

OECD (1999). *The Knowledge-Based Economy: A Set of Facts and Figures*, Paris: OECD.

OECD (2000). "Business-to-Business Electronic Commerce in Publishing, Retail Distribution and Pharmaceuticals Distribution in France", DSTI/ICCP/IE(99)9/FINAL, Paris.

OECD (2001a). "The New Economy: Beyond the Hype – The OECD Growth Project", Paris: OECD.

OECD (2001b). "Science, Technology and Industry Outlook – Drivers of Growth: Information Technology, Innovation and Entrepreneurship", Paris: OECD.

OECD, (2001c). "Innovative Networks: Cooperation in National Innovation Systems", Paris: OECD.

OECD (2002a, forthcoming). "Information Technology Outlook", Paris: OECD.

OECD (2002b). "EBIP Background Documents (Electronic Commerce Business Impacts Project, Sector and National Reports), (available at: http://www.oecd.org/oecd/pages/home/ displaygeneral/0,3380,EN-document-29-nodirectorate-no-20-20570-29,FF.html).

OECD (2002c forthcoming). "The Impacts of Electronic Commerce on Business", Paris: OECD.

OECD (2002d). "The Working Party on Indicators for the Information Society (WPIIS): Progress Report and Ideas for Future Work", DSTI/ICCP(2002)4.

Oliver, A.L. (2001). "Strategic Alliances and the Learning Life-Cycle of Biotechnology Firms", *Organization Studies*, Vol. 22, No. 3, pp. 467-90.

Oliver, A.L. (1994). "In Between Markets and Hierarchies – Networking through the Life Cycle of New Biotechnology Firms", Institute of Social Science Research Working Paper, UCLA.

Oliver, R. (1999). *The Coming Biotech Age: The Business of Bio-Materials*, New York: McGraw-Hill.

Orscnigo, L., Pammolli, F., Riccaboni, M., Bonaccorsi, A. and Turchetti, G. (1998). "The Evolution of Knowledge and the Dynamics of Industry Networks", *Journal of Management and Governance*, Vol. 1, pp. 147-75.

Orsenigo, L., Pammolli, F. and Riccaboni, M. (2001). "Technological Change and Network Dynamics. Lessons from the Pharmaceutical Industry", *Research Policy*, Vol. 30, pp. 485-508.

Osberg, L. (2001). "Needs and Wants: What is Social Progress and How Should it be Measured?", in Banting, K., Sharpe, A. and St.-Hilaire, F. (eds.), *The Review of Economic Performance and Social Progress, The Longest Decade: Canada in the 1990s*, Montreal and Ottawa: The Institute for Research on Public Policy and Centre for the Study of Living Standards, pp. 23-41.

Osborn, R.N. and Hagedoorn, J. (1997). "The Institutionalization and Evolutionary Dynamics of Interorganizational Alliances and Networks", *Academy of Management Journal*, Vol. 40, No. 2, pp. 261-78.

Pal, L.A. (2001). *Beyond Policy Analysis: Public Issue Management in Turbulent Times*, Toronto: Nelson.

Pavitt, K. (1984). "Sectoral Patterns of Technical Change: Towards a Taxonomy and a Theory", *Research Policy*, Vol. 13, No. 6, pp. 343-373.

Pavitt, K. (1988). "Uses and Abuses of Patent Statistics", in van Raan, A.F.J. (ed.), *Handbook of Quantitative Studies of Science and Technologies*, Elsevier Science Publishers.

Pavitt, K. (1991). "What Makes Basic Research Economically Useful?", *Research Policy*, Vol. 20, pp. 109-119.

Pavitt, K. (1998). "Technologies, Products and Organization in the Innovating Firm: What Adam Smith Tells Us and Joseph Schumpeter Doesn't", *Industrial and Corporate Change*, Vol. 7 No. 3, pp. 433-452.

Pavitt, K. (2000). "Knowledge Specialization, Organization and the Boundaries of the Firm", BPRC Conference on Knowledge Management: Concepts and Controversies, Warwick, 10-11 February.

Pavitt, K. and Steinmueller, W.E. (2002). "Technology and Corporate Strategy: Change, Continuity and the Information Revolution", in Pettigrew, A., Thomas, H. and Whittington, R. (eds.), *Handbook of Strategy and Management*, London: Sage Publications.

Penrose, E.T. (1959). *Theory of the Growth of the Firm*. Oxford: Blackwell.

Piore, M. and Sabel, C. (1984). *The Second Industrial Divide*, New York: Basic Books.

Pisano, G.P. (1990). "The R&D Boundaries of the Firm: An Empirical Analysis", *Administrative Science Quarterly*, Vol. 35, pp. 153 -76.

Polanyi, M. (1966). *The Tacit Dimension*, Garden City, NY: Doubleday.

Popper, S.W. and Wagner, C.S. (2001). "New Foundations for Growth: The U.S. Innovation System Today and Tomorrow", RAND Science and Technology Institute, Washington, D.C.

Posner, M. (2002). "Climate of Fear is Putting Our Freedoms at Risk", *The Globe and Mail*, 20 February, B15.

Powell, W.W. (1990). "Neither Market nor Hierarchy: Network Forms of Organization", *Research in Organizational Behavior*, Vol. 12, pp. 295-336.

Powell, W.W., Koput, K.W. and Smith-Doerr, L. (1996). "Interorganisational Collaboration and the Locus of Innovation: Networks of Learning in Biotechnology", *Administrative Science Quarterly*, Vol. 41, pp. 116-45.

Quintas, P. (2002). "Managing Knowledge in a New Century", in Little, S., Quintas, P. and Ray, T. (eds.), *Managing Knowledge*, London: Sage Publications.

Quintas, P. and Guy, K. (1989). "Alvey in Industry: Corporate Strategy and the Alvey Programme", Report to the Department of Trade and Industry, Science Policy Research Unit, Falmer, Sussex UK.

Quintas, P. and Guy, K. (1995). "Collaborative, Pre-competitive R&D and the Firm", *Research Policy*, Vol. 24, pp. 325-348.

Rabinow, P. (1996). *Making PCR: A Story of Biotechnology*, Chicago: University of Chicago Press.

Rae, G. (1999). "*Collective Invention in Software in Insurance* ", Colline WP No. 10, IMRI, Paris: Université de Paris Dauphine.

Rahm, D. (1995). "University-Firm Linkages for Industrial Innovation, Final Report to the National Science Foundation", Grant No. SBR 93-05591, University of South Florida, Tampa.

Rice, J.J. and Prince, M.J. (2000). *Changing Politics of Canadian Social Policy*, Toronto: University of Toronto Press.

Richardson, G.B. (1972). "The Organization of Industry", *Economic Journal*, Vol. 82, pp. 883-96.

Robson, M., Townsend, J. and Pavitt, K. (1988). "Sectoral Patterns of Production and Use of Innovation in UK: 1945-1983", *Research Policy*, Vol. 17, pp. 1-14.

Rosenberg, N. and Birdzell, L. (1990). "Science, Technology and the Western Miracle", *Scientific American*, Vol. 263, No. 5.

Rothaermel, F.T. (2001). "Complementary Assets, Strategic Alliances, and the Incumbent's Advantage: an Empirical Study of Industry and Firm Effects in the Biopharmaceutical Industry", *Research Policy*, Vol. 30, No. 8, pp. 1235-1252.

Rothwell, R. and Zegveld, W. (1984). *Industrial Innovation and Public Policy*, London: Pinter.

Rycroft, R.W. and Kash, D.E. (1999). *The Complexity Challenge: Technological Innovation for the 21st Century*, London: Pinter.

Sachwald, F. (1998). "Cooperative Agreements and the Theory of the Firm: Focusing on Barriers to Change", *Journal of Economic Behavior and Organization*, Vol. 35, pp. 203-225.

Sahal, D. (1981). "The Farm Tractor and the Nature of Technological Innovation", *Research Policy*, Vol. 10, pp. 368-402.

Saviotti, P.P. (1988). "The Measurement of Change in Technological Output," A. F. J. van Raan, *Handbook of Quantitative Studies of Science and Technologies*, London: Elsevier Science Publishers, pp. 555-611.

Schaan, S. and Anderson, F. (2001). "Innovation in Canadian Manufacturing: National Estimates", Statistics Canada Cat. No. 88F006XIE, SIEID Working Paper #10, pp. 39-40, Ottawa: Statistics Canada.

Schumpeter, J. (1947). *Capitalism, Socialism and Democracy,* (Second Edition), New York: Harper and Row.

Senker, J. (1995). "Tacit Knowledge and Models of Innovation", *Industrial and Corporate Change*, Vol. 2, pp. 425-447.

Senker, J. and Sharp, M. (1997). "Organizational Learning in Cooperative Alliances: Some Case Studies in Biotechnology", *Technology Analysis and Strategic Management*, Vol. 9, No. 1, pp. 35-51.

Sharp, M. (1996). "Knowledge Transfer and Biotechnology: The Case of Europe's Large Multinationals", in Inzelt, A. and Coenen, R. (eds.), *Knowledge, Technology Transfer and Foresight*, Boston: Kluwer.

Silver, M. (1984). *Enterprise and the Scope of the Firm*, London: Martin Robertson.

Sing, K. (1997). "The Impact of Technological Complexity and Interfirm Cooperation on Business Survival", *Academy of Management Journal*, Vol. 40, No. 2, pp. 339-67.

Smith, A. (1887). *An Inquiry into the Nature and Causes of the Wealth of Nations*, London: George Bell & Sons.

Smith, A. (1976 [1776]). "An Inquiry into the Nature and Causes of the Wealth of Nations", in Campbell, R.H., Skinner, A.S. and Todd, W.B. (eds.), 2 Volumes, Oxford: Oxford University Press.

Spender, J.C. (1996). "Making Knowledge the Basis of a Dynamic Theory of the Firm", *Strategic Management Journal*, Vol. 17, pp. 45-62, Winter.

Stanley, D. (2001). "Communities: The Engines of Social Cohesion", *Policy Research Initiative Horizons*, Vol. 4, No. 5, pp. 4-6.

Statistics Canada (1998). *Science and Technology Activities and Impacts: A Framework for a Statistical Information System*, Statistics Canada catalogue no. 88-522, Ottawa: Statistics Canada.

Stehr, N. (1994). *Knowledge Societies*, London: Sage.

Steier, L. and Greenwood, R. (2000). "Entrepreneurship and the Evolution of Angel Financial Networks", *Organization Studies*, Vol. 21, No. 1, pp. 163-92.

Stewart, A.T. (1998). "The Price of Ideas", in Herzberg, A.M. and Krupka, I. (eds.), *Statistics, Science and Public Policy*, Kingston, Ontario: Queen's University, pp. 119-124.

Stigler, G.J. (1951). "The Division of Labor is Limited by the Extent of the Market", *Journal of Political Economy*, Vol. 59, No. 3, pp. 185-93.

Stoneman, P. (1995). *Handbook of the Economics of Innovation and Technological Change*, Oxford: Blackwell.

Storper, M. (1996). "Innovation as Collective Action: Conventions, Products and Technologies", *Industrial and Corporate Change*, Vol. 5, No. 3, pp. 761-790.

Stuart, T.E. (2000). "Interorganisational Alliances and the Performance of Firms: A Study of Growth and Innovation Rates in a High-Technology Industry", *Strategic Management Journal*, Vol. 21, pp. 791-811.

Stuart, T.E., Hoang, H. and Hybels, R.C. (1999). "Interorganizational Endorsements and the Performance of Entrepreneurial Ventures", *Administrative Science Quarterly*, Vol. 44, No. 2, pp. 315-349.

Swann, G.M.P. and Prevezer, M. (1998). "Introduction", in Swann, G.M.P. and Prevezer, M. and Stout, D. (eds.), *The Dynamics of Industrial Clustering*, Oxford: Oxford University Press.

Taylor, P., Quintas, P., Storey, J. and Fowle, W. (2000). "Utilising Internal and External Resources for Innovation: Employment Practices and Inter-Firm Collaboration", Working Paper 2, MKIRU, Milton Keynes: Open University Business School.

Teece, D.J. (1976). *The Multinational Corporation and the Resource Cost of International Technology Transfer*, Cambridge, MA: Ballinger.

Teece, D.J. (1981). "The Market for Know-How and the Efficient International Transfer of Technology', *The Annals of the American Academy of Political and Social Science*, Vol. 458, pp. 81-96.

Teece, D.J. (1986). "Profiting from Technological Innovation: Implications for Integration, Collaboration, Licensing and Public Policy", *Research Policy*, Vol. 15, No. 6, pp. 285-305.

Teece, D.J. (1990). "Capturing Value Through Cooperative Competition", in de la Mothe, J. and Ducharme, L.-M. (eds.), *Science, Technology and Free Trade*, London: Pinter.

Thomson, G. (1954). "Intercompany Technical Standardisation in the Early American Automobile Industry", *The Journal of Economic History*, Vol. 14, No. 1.

Thomson, W. (1889). *Popular Letters and Addresses, Volume 1*, London: Macmillan.

Tsoukas, H. (1992). "Ways of Seeing – Topographic and Network Representations in Organization Theory", *Systems Practice*, Vol. 5, No. 4, pp. 441-56.

Tushman, M. and Anderson, P. (1986). "Technological Discontinuities and Organizational Environments", *Administrative Science Quarterly*, Vol. 31, pp. 439-65.

Tzokas, N. and Saren, M. (1997). "On Strategy, Typologies and the Adoption of Technological Innovations in Industrial Markets", *British Journal of Management*, Vol. 8, S91-S105, Special Issue.

U.S. Government Accounting Office (GAO) (1994). "Technology Transfers: Benefits of Cooperative R&D Agreements", Washington, DC: U.S. General Accounting Office.

U.S. Office of Technology Policy (1996). "Effective Partnering: A Report to Congress on Federal Technology Partnerships", Washington, DC: U.S. Department of Commerce.

United Nations (2000). *1999 International Trade Statistics Yearbook, Volume I: Trade by Country* and *Volume II: Trade by Commodity*, New York: United Nations.

Utterback, J.M. (1994). *Mastering the Dynamics of Innovation: How Companies Can Seize Opportunities in the Face of Technological Change*, Cambridge, MA: Harvard Business School Press.

Utterback, J.M. and Abernathy, W.J. (1975). "A Dynamic Model of Process and Product Innovation", *Omega*, Vol. 3, No. 6, pp. 639-56.

Veblen, T. (1939 [1915]). *Imperial Germany and the Industrial Revolution*, New York: Viking Press.

Veblen, T. (1992 [1899]). *The Theory of the Leisure Class: An Economic Study of Institutions*, New Brunswick, NJ: Transaction Publishers.

Vermeulen, F. and Barkema, H. (2001). "Learning through Acquisition", *Academy of Management Journal*, Vol. 44, No. 3, pp. 457-76.

Veugelers, R. (1997). "Internal R&D Expenditures and External Technology Sourcing", *Research Policy*, Vol. 26, pp. 303-315.

Veugelers, R. and Cassiman, B. (1999). "Make and Buy-In Innovation Strategies: Evidence from Belgian Manufacturing Firms", *Research Policy*, Vol. 28, pp. 63-80.

Vonortas, N. (1997a). "Research Joint Ventures in the U.S.", *Research Policy*, Vol. 26, No. 4-5, pp. 577-595.

Vonortas, N. (1997b). "The Process of Technological Innovation", in Vonortas, N. (ed.), *Cooperation in Research and Development*, Boston: Kluwer.

Wajcman, J. (2002). "Addressing Technological Change: The Challenge to Social Theory", *Current Sociology*, Vol. 50, No. 3.

Ward, L. (1906). *Applied Sociology: A Treatise on the Conscious Improvement of Society by Society*, Boston: Ginn and Company.

Wellman, B. (2001). "Computer Networks as Social Networks", *Science*, Vol. 293, pp. 2031-2034.

Wenger, E. (1998). *Communities of Practice*, Cambridge, UK: Cambridge University Press.

Wenger, E. and Snyder, W. (2000). "Communities of Practice: The Organizational Frontier", *Harvard Business Review*, January – February.

Williamson, O.E. (1975). *Markets and Hierarchies: Analysis and Antitrust Implications*, London: The Free Press.

Williamson, O.E. (1983). "Organizational Innovation: The Transaction-Cost Approach", in Ronen, J. (ed.), *Entrepreneurship*, Lexington, MA: Heath Lexington, pp. 101-134.

Winter, S. (1984). "Schumpeterian Competition in Alternative Technological Regimes", *Journal of Economic Behavior and Organization*, Vol. 5, No. 3-4, pp. 287-320.

Young, A. (1928). "Increasing Returns and Economic Progress", *Economic Journal*, Vol. 38, pp. 527-42.

Zucker, L., Darby, M., Brewer, M. and Peng, Y. (1995). *Collaboration Structure and Information Dilemmas In Biotechnology: Organizational Boundaries As Trust Production*. Cambridge: National Bureau of Economic Research.

Zucker, L.G., Darby, M.R. and Peng, Y. (1998). *Fundamentals of Population Dynamics and The Geographic Distribution of U.S. Biotechnology Enterprises, 1976-1989*, Cambridge: National Bureau of Economic Research.

Zucker, L.G., Darby, M.R. and Armstrong, J. (2001). *Commercializing Knowledge: University Science, Knowledge Capture, and Firm Performance in Biotechnology*, Cambridge, MA: National Bureau of Economic Research.

INDEX

H

I

J

K

L

M

N

O

P

R

S

T

U

1. A. Phillips, A.P. Phillips and T.R. Phillips:
   Biz Jets. Technology and Market Structure in the Corporate Jet Aircraft Industry. 1994 ISBN 0-7923-2660-1
2. M.P. Feldman:
   The Geography of Innovation. 1994 ISBN 0-7923-2698-9
3. C. Antonelli:
   The Economics of Localized Technological Change and Industrial Dynamics. 1995 ISBN 0-7923-2910-4
4. G. Becher and S. Kuhlmann (eds.):
   Evaluation of Technology Policy Programmes in Germany. 1995 ISBN 0-7923-3115-X
5. B. Carlsson (ed.): *Technological Systems and Economic Performance: The Case of Factory Automation.* 1995 ISBN 0-7923-3512-0
6. G.E. Flueckiger: *Control, Information, and Technological Change.* 1995 ISBN 0-7923-3667-4
7. M. Teubal, D. Foray, M. Justman and E. Zuscovitch (eds.):
   Technological Infrastructure Policy. An International Perspective. 1996 ISBN 0-7923-3835-9
8. G. Eliasson:
   Firm Objectives, Controls and Organization. The Use of Information and the Transfer of Knowledge within the Firm. 1996 ISBN 0-7923-3870-7
9. X. Vence-Deza and J.S. Metcalfe (eds.):
   Wealth from Diversity. Innovation, Structural Change and Finance for Regional Development in Europe. 1996 ISBN 0-7923-4115-5
10. B. Carlsson (ed.):
    Technological Systems and Industrial Dynamics. 1997 ISBN 0-7923-9940-4
11. N.S. Vonortas:
    Cooperation in Research and Development. 1997 ISBN 0-7923-8042-8
12. P. Braunerhjelm and K. Ekholm (eds.):
    The Geography of Multinational Firms. 1998 ISBN 0-7923-8133-5
13. A. Varga:
    University Research and Regional Innovation: A Spatial Econometric Analysis of Academic Technology Transfers. 1998 ISBN 0-7923-8248-X
14. J. de la Mothe and G. Paquet (eds.):
    Local and Regional Systems of Innovation, 1998 ISBN 0-7923-8287-0
15. D. Gerbarg (ed.):
    The Economics, Technology and Content of Digital T V, 1999 ISBN 0-7923-8325-7
16. C. Edquist, L. Hommen and L. Tsipouri
    Public Technology Procurement and Innovation, 1999 ISBN 0-7923-8685-X
17. J. de la Mothe and G. Paquet (eds.):
    Information, Innovation and Impacts, 1999 ISBN 0-7923-8692-2

18. J. S. Metcalfe and I. Miles (eds.):
Innovation Systems in the Service Economy: Measurement and Case Study Analysis, 2000 ISBN 0-7923-7730-3
19. R. Svensson:
Success Strategies and Knowledge Transfer in Cross-Border Consulting Operations, 2000 ISBN 0-7923-7776-1
20. P. Braunerhjelm:
Knowledge Capital and the "New Economy": Firm Size, Performance and Network Production, 2000 ISBN 0-7923-7801-6
21. J. de la Mothe and J. Niosi (eds.):
The Economic and Social Dynamics of Biotechnology, 2000 ISBN 0-7923-7922-5
22. B. Guilhon, (ed.):
Technology and Markets for Knowledge: Knowledge Creation, Diffusion and Exchange within a Growing Economy, 2000 ISBN 0-7923-7202-6
23. M. Feldman and A. Link (eds.):
Innovation Policy in the Knowledge-Based Economy, 2001 ISBN 0-7923-7296-4
24. J. de la Mothe and D. Foray (eds.):
Knowledge Management in the Innovation Process 2001 ISBN 0-7923-7464-9
25. M. Feldman and N. Massard (eds.):
Institutions and Systems in the Geography of Innovation, 2002 ISBN: 0-7923-7614-5
26. B. Carlsson (ed.):
Tehnological Systems in the Bio Industries: An International Study, 2002 ISBN: 0-7923-7633-1
27. D. Audretsch, R. Thurik, I. Verheul, and S. Wennekers (eds.): *Entrepreneurship: Determinants and Policy in a European-US Comparison*, 2002 ISBN: 0-7923-7685-4
28. J. de la Mothe and A. Link (eds.):
Networks, Alliances and Partnerships in the Innovation Process ISBN: 1-4020-7172-8

KLUWER ACADEMIC PUBLISHERS – BOSTON / DORDRECHT / LONDON